Integrated Series in Information Systems

Volume 34

For further volumes:
http://www.springer.com/series/6157

Kweku-Muata Osei-Bryson
Ojelanki Ngwenyama
Editors

Advances in Research Methods for Information Systems Research

Data Mining, Data Envelopment Analysis, Value Focused Thinking

Springer

Editors
Kweku-Muata Osei-Bryson
Department of Information Systems
Virginia Commonwealth University
Richmond, VA
USA

Ojelanki Ngwenyama
Department of Information Systems
Ryerson University
Toronto, ON
Canada

ISSN 1571-0270 ISSN 2197-7968 (electronic)
ISBN 978-1-4614-9462-1 ISBN 978-1-4614-9463-8 (eBook)
DOI 10.1007/978-1-4614-9463-8
Springer New York Heidelberg Dordrecht London

Library of Congress Control Number: 2013953894

Printed on acid-free paper

Springer is part of Springer Science+Business Media (www.springer.com)

Contents

Contributors

Kwasi Amoako-Gyampah Bryan School of Business and Economics, The University of North Carolina at Greensboro, Greensboro, NC 27402-6170, USA, e-mail: k_amoako@uncg.edu

Francis Kofi Andoh-Baidoo Department of Computer Information Systems and Quantitative Methods, The University of Texas—Pan American, Edinburg, TX 78539, USA, e-mail: andohbaidoof@utpa.edu

Arlene Bailey Department of Sociology, The University of the West Indies, Mona, Kingston 7, Jamaica, e-mail: Arlene.Bailey@UWIMona.Edu.JM

Corlane Barclay School of Computing and Information Technology, University of Technology, 237 Old Hope Road, Kingston, Jamaica, e-mail: cbarclay@utech.edu.jm

Felix Bollou School of Information Technology and Communications, American University of Nigeria, Yola 640001, Adamawa, Nigeria, e-mail: bollou@gmail.com

Myung Ko Department of Information Systems and Technology Management One UTSA Circle, The University of Texas at San Antonio, San Antonio, TX 78249, USA, e-mail: Myung.Ko@utsa.edu

Yan Li Department of Information Systems, Virginia Commonwealth University, 301 W. Main Street, Richmond, VA 23284, USA, e-mail: LIY26@VCU.Edu

Ojelanki Ngwenyama Ted Rogers School of Management, Ryerson University, 350 Victoria Street, Toronto, ON M5B 2K3, Canada, e-mail: Ojelanki@ryerson.ca

Kweku-Muata Osei-Bryson Department of Information Systems, Virginia Commonwealth University, 301 W. Main Street, Richmond, VA 23284, USA, e-mail: KMOsei@VCU.Edu

Sergey Samoilenko Department of Computer Science, Averett University, 420 W Main St Danville, VA 24541, USA, e-mail: SSamoilenko@Averett.Edu

Manoj Thomas Department of Information Systems, Virginia Commonwealth University, 301 W. Main Street, Richmond, VA 23284, USA, e-mail: mthomas@vcu.edu

Chapter 1
Introduction

Kweku-Muata Osei-Bryson and Ojelanki Ngwenyama

The decades from the 1990s to 2000s have seen long and vigorous debates over the perceived deepening alienation of the academic discipline of information systems from practice of and technical content of information systems (Iiviri 2003; Robey 1996; Markus 1997; Robey and Markus 1998; Orlikowski and Iacono 2001; Weber 2006). While during the same period, the quantity and variety of research outputs increased, the influence of IS research in business organizations declined. Some have suggested that the focus of IS research had become too narrow and outputs of this research to foreign to IS practitioners and organizational managers (Markus 1997; Hirschheim and Klein 2006). Reflecting on this issue, some researchers point to, among other factors, the rise of positivist approaches focusing on rigor and the decline of pragmatic approaches focusing on relevance as a primary cause of the alienation (Ciborra 1998; Hirschheim and Klein 2000, 2006).

Out of this debate have come many prescriptions for "fixing" the crisis (cf. Agarwal and Lucas 2005; Benbasat and Zmud 2003; Lyytinen and King 2006; Taylor et al. 2010). Our interest as researchers is how to harness information technology advancements to support non-traditional post-positivist research methods which could enhance our research practice, contribute to the production of IS knowledge that is useful to a broader set of stakeholders, and bridge the communication gap between IS researchers and key stakeholders. The post-positivist philosophies of social science such as critical realism, sociomateriality, and critical

K.-M. Osei-Bryson (✉)
Virginia Commonwealth University, Department of Information Systems,
301 W. Main Street, Richmond, VA 23284, USA
e-mail: KMOsei@VCU.Edu

O. Ngwenyama
Ryerson University, Ted Rogers School of Management, 350 Victoria Street, Toronto, ON
M5B 2K3, Canada
e-mail: Ojelanki@ryerson.ca

K.-M. Osei-Bryson and O. Ngwenyama (eds.), *Advances in Research Methods for Information Systems Research*, Integrated Series in Information Systems 34,
DOI: 10.1007/978-1-4614-9463-8_1,

social theory have exposed the fundamental limitations of the positivist behavioral approach to IS research and offer new frontiers for the systematic development of new scientific research practice within the discipline of IS (Mingers 2003, 2004; Smith 2006; Myers and Klein 2011). They also suggest the need to explore other non-traditional IS research paths.

Our exploration of quantitative, non-traditional IS research paths began several years ago with the use of data mining (DM) methods for data analysis. DM methods aim to identify non-trivial, interesting patterns that are embedded in a given dataset. Unlike confirmatory approaches, DM involves "working up from the data" and as such can result in the discovery of new knowledge which is one of the aims of scientific inquiry. Results from that initial exploration were encouraging and motivated us to explore the use of DM and other techniques in other areas of our own research.

In this book, we demonstrate how developments in software technologies for DM such as regression splines and decision tree induction could assist researchers in systematic post-positivist theory testing and development and also to provide answers to some research questions that cannot be addressed using traditional approaches. We also demonstrate how some established management science techniques such as data envelopment analysis (DEA) and value-focused thinking (VFT) can be used in combination with traditional statistical analysis approaches (e.g., structural equation modeling) and/or with DM approaches (e.g., clustering, decision tree induction, regression splines) to more effectively explore IS behavioral research questions. We also demonstrate how these techniques can be combined and used in multi-method research strategies to address a range of empirical problems in information systems. Further, some of these techniques (e.g., VFT) can also be used in IS design science research.

It is also important to note that while these techniques have resulted from research in the IS and management science communities and studies that applied them to IS research problems have appeared in reputable journals in both fields, it would appear that the vast majority of IS doctoral students have not been exposed to them as part of their research methodology coursework. Further, many experienced IS researchers are also insufficiently familiar with these techniques and potential uses. One major reason for this situation is that there is no single book that provides information and guidance to doctoral students and more advanced researchers on the use of these techniques for IS and other business-related researchers.

In this book, we focus on DM, DEA, and VFT. We are not claiming that these are the only non-traditional research techniques that are relevant. However, it would take a huge book to cover all such techniques, and the size of such a book might be an intimidating turn-off to many novice and advanced researchers. Further, we do not claim that we have covered all of the possible use of these techniques as components of new research methods that could add value to IS research. This book fills a gap in the training and development of IS and other researchers and provides motivation for the exploration and use of other non-traditional research methods. This could be the first in a series of such value-adding books for the IS, other business and social science research communities as well as practitioners.

Academic researchers, including experienced researchers and doctoral students, engaged in behavioral science research, particularly in the area of information systems, should find the material in this book to be useful as a resource on research methods in information systems. The methods presented are also applicable to researchers in other areas of business and the social sciences. They could also be used in design science research to develop method artifacts. Examples of such use are included. This book can also be used by business practitioners to understand organizational phenomena and support improved decision-making.

The rest of the book is as organized as follows:

- Chapter 2 presents a short overview of the fundamental inferential logics of inquiry upon positivist and post-positivist social science inquiry methods have been developed. Its aim is to facilitate a conceptual understanding of the underlying inferential mechanisms of the different approaches to scientific inquiry illustrated in this book.
- The next eight (8) chapters focus on the DM modeling techniques, with four focusing on applications of selected DM modeling techniques, one (1) involving a manual extraction of decision trees from qualitative data, and the other three (3) providing overviews on the selected DM modeling techniques.
  - Chapter 3 ("Overview on Decision Tree Induction") provides an overview of decision tree induction. Its main purpose is to introduce the reader to the major concepts underlying this DM technique.
  - Chapter 4 ("Using Decision Tree Induction for Theory Development") explores and illustrates how DM techniques could be applied to assist researchers in systematic theory testing and development. It presents an approach that involves the use of the decision tree modeling and traditional statistical hypothesis testing to automatically abduct hypotheses from data.
  - Chapter 5 ("A Hybrid Decision Tree-based Method for Exploring Cumulative Abnormal Returns") presents a hybrid method for investigating the capital markets reaction to the public announcement of a business-related event (e.g., a security breach). The hybrid method involves the application of the event study methodology, decision tree generation (DT), together with statistical hypothesis testing.
  - Chapter 6 ("An Ethnographic Decision Tree Modeling: An Exploration of Telecentre Usage in the Human Development Context") presents the use of decision tree modeling in an investigation of the decision-making process of community members who decide on using telecenters to support economic livelihood. Unlike the other papers that involve decision trees, the generation of the decision trees in this paper did not involve the use of DM software but rather involved extracting
  - Chapter 7 ("Using Association Rules Mining To Facilitate Qualitative Data Analysis in Theory Building") involves the use of the association rules induction technique as a major component of a new procedure that aims to facilitate the development of propositions/hypotheses from qualitative data. The proposed procedure is illustrated using a case study from the public health domain.

- Chapter 8 ("Overview on Multivariate Adaptive Regression Splines") provides an overview of multivariate adaptive regression splines. Its main purpose is to introduce the reader to the major concepts underlying this DM technique, particularly those that are relevant to the chapter that involves the use of this technique.
- Chapter 9 ("Reexamining the Impact of Information Technology Investment on Productivity Using Regression Tree and MARS") involves the use of multiple DM techniques to explore the issue of the impact of investments in IT on productivity. The use of this pair of DM techniques allowed for the exploration of interactions between the input variables as well as conditional impacts.
- Chapter 10 ("Overview on Cluster Analysis") provides an overview of cluster analysis. Its main purpose is to introduce the reader to the major concepts underlying this DM technique, particularly those that are relevant to the chapter that involves the use of this technique.

• The next three chapters focus on the data envelopment analysis technique:
  - Chapter 11 ("Overview on Data Envelopment Analysis") provides overview of DEA. Its main purpose is to introduce the reader to the major concepts underlying this nonparametric technique. It also discusses previous applications of DEA in information systems research.
  - Chapter 12 ("Exploring the ICT Utilization using DEA") presents a DEA-based methodology for assessing the efficiency of investments in ICT. Measuring the efficiency of investments in ICT infrastructure could provide insights relevant for effective allocation of scarce resources in developing countries. The analysis uses statistical data on the ICT sectors of six West African countries.
  - Chapter 13 ("A DEA-centric Decision Support System for Monitoring Efficiency-Based Performance") describes an organizational decision support system (DSS) that aims to address some of the challenges facing organizations competing in dynamic business environments. This DSS utilizes DEA and several DM methods.

• The final two chapters focus on the VFT methodology.
  - Chapter 14 ("Overview on the Value Focused Thinking Methodology") provides an overview of the VFT methodology. Its main purpose is to introduce the reader to the major concepts of this methodology, particularly those that are relevant to the next chapter. It also discusses previous applications of the VFT methodology in information systems research.
  - Chapter 15 ("Using Value Focused Thinking to Develop Performance Criteria and Measures for Information Systems Projects") addresses the issue of selecting project performance criteria that reflect the values of the relevant project stakeholders. A hybrid method for addressing this issue is presented. It relies on the principles and advantages of the VFT and goal–question–metric (GQM) methods. A case study is used to illustrate and assess the hybrid method.

References

Agarwal R, Lucas HC Jr (2005) The information systems identity crisis: focusing on high-visibility and high-impact research. MIS Q 29(3):381–398

Benbasat I, Zmud RW (2003) The identity crisis within the IS discipline: defining and communicating the discipline's core properties. MIS Q 27(2):183–194

Ciborra CU (1998) Crisis and foundations: an inquiry into the nature and limits of models and methods in the information systems discipline. J Strateg Inf Syst 7(1):5–16

Hirschheim RA, Klein HK (2006) Crisis in the IS field? A critical reflection on the state of the discipline. In: King JL, Lyytinen K (eds) Information systems: the state of the field. Wiley, Chichester, pp 71–146

Iivari J (2003) The IS core-VII towards information systems as a science of meta-artifacts. Commun Assoc Inf Syst (Volume 12, 2002):568, 581

Lyytinen K, King JL (2006) Nothing at the center?: Academic legitimacy in the information systems field. Inf Syst: State Field 5(6):233–266

Markus ML (1997) The qualitative difference in information systems research and practice. In: Lee AS, Liebenau J, DeGross JI (eds) Information systems and qualitative research. Springer, London, US, pp 11–27

Mingers J (2003) The paucity of multimethod research: a review of the information systems literature. Inf Syst J 13(3):233–249

Mingers J (2004) Realizing information systems: critical realism as an underpinning philosophy for information systems. Inf Organ 14(2):87–103

Myers MD, Klein HK (2011) A set of principles for conducting critical research in information systems. MIS Q 35(1):17–36

Orlikowski WJ, Iacono CS (2001) Research commentary: desperately seeking the" it" in it research—a call to theorizing the it artifact. Inf Syst Res 12(2):121–134

Robey D, Markus ML (1998) Beyond rigor and relevance: producing consumable research about information systems. Inf Res Manage J (IRMJ) 11(1):7–16

Robey D (1996) Research commentary: diversity in information systems research: threat, promise, and responsibility. Inf Syst Res 7(4):400–408

Smith ML (2006) Overcoming theory-practice inconsistencies: Critical realism and information systems research. Inf Organ 16(3):191–211

Taylor H, Dillon S, Van Wingen M (2010) Focus and diversity in information systems research: meeting the dual demands of a healthy applied discipline. MIS Q 34(4):647–667

Weber R (2006) Still desperately seeking the IT artifact. Inf Syst: State Field 7(4):43–55

Chapter 2
Logical Foundations of Social Science Research

Ojelanki Ngwenyama

In this chapter, I want to review the four inferential logics (1) induction, (2) deduction, (3) abduction, and (4) retroduction which we use to develop the conjectures or hypotheses when doing theory development. The reason for this review is to provide a conceptual understanding of the underlying inferential mechanisms of the different approaches to scientific inquiry illustrated in this book. While we have always been using all four types of inferential logic, there is still misunderstanding of data analytic methods applying some of the four basic inferential mechanisms can contribute to the development of scientific theories.

1 Introduction

Social science inquiry is founded on the idea of achieving self-understanding (Winch 1958). Social science inquiry is essentially an inquiry into ourselves, our agency as human actors, and the social world we create (Berger and Luckmann 1991). While there are various approaches to social science inquiry which hold different positions on the nature of our social world, its structures and participants, the underlying logics of social inquiry are fundamentally the same. Why so? Because our logic systems do not deal in reality; they use symbols (representations of reality) and logical rules for reasoning about reality. Furthermore, all our scientific theories are based on representations of reality. Even our "empirical observations" are nothing more than interpretations of our perceptions of reality. This chapter, I want

O. Ngwenyama (✉)
Ryerson University, Ted Rogers School of Management, 350 Victoria Street, Toronto, ON M5B 2K3, Canada
e-mail: Ojelanki@ryerson.ca

O. Ngwenyama
Faculty of Commerce, University of Cape Town, CapeTown, South Africa

K.-M. Osei-Bryson and O. Ngwenyama (eds.), *Advances in Research Methods for Information Systems Research*, Integrated Series in Information Systems 34,
DOI: 10.1007/978-1-4614-9463-8_2, © Springer Science+Business Media New York 2014

to revisit the fundamental logics that we use to manipulate symbols upon which we construct our scientific theories about reality. Theories are a substitute for direct knowing. As Popper argued, a theory is an unjustified statement, a probable explanation, a conjecture or hypothesis about reality. Our theories are simply claims that we make about reality (Toulmin and Barzun 1981). We attempt to justify our claims (theories) by subjecting them to logical criticism and empirical testing. Of course, neither logical criticism nor testing can reveal or validate truth content, they can only test consistency or conformity with a predefined set of rules (Carnap 1953) and correspondence with empirical observations (Popper 1972). It must also be noted that the symbols we use to represent reality are not reality, *ergo*, they are interpretations that carry no truth content. We can interrogate how well out claims are argued or how consistent they are with existing claims that have remained non-rejected in our system of knowledge, but we cannot assert their truth (Hempel and Oppenheim 1965). We can also interrogate the implications of our theories and their ability to predict some specific observations of reality. In this way, we subject our theories to criticism, explore their plausibility to explain social reality, and when we arrive at a better explanation we abandon them (Harman 1965, Popper 1959, Thagard 1978). It is also important to note here that social theories are time bound and culture sensitive, because human action is based on beliefs, and belief systems change over time, whatever the rate of change.

2 Empirical-Based Social Science Inquiry

The general model (Table 1) of empirically based social science inquiry starts when the scientist observes some puzzling phenomena or phenomenal behavior in the universe of inquiry. Taking in particular information using the senses, he or she

Table 1 The general model of empirically based social science inquiry

Phase	Activity	Outcome
Empirical observation	Gather data about some phenomena or phenomenal behavior of interest	Identification of some puzzle about the phenomena or phenomenal behavior to be solved
Hypotheses generation	Invent one or more conjectures or hypotheses to explain the phenomena or phenomenal behavior	Hypotheses to be examined
Design experiments	Design observation experiments to test the logical consequences of the hypotheses	Observation experiments of the form "if principle P is true, then event E should occur or fact F should be true"
Empirical testing	Collect observations about the phenomena and examine them to see if the predictions prove to be true or false Test also the reliability of the event E occurring when P is true or the F is observed when P is true	A rejected or non-rejected theoretical explanation of the phenomena or phenomenal behavior and some reliability measure

then attempts to conjecture some probable explanations (hypotheses or propositions) of the phenomena or phenomenal behavior. It is in this step of hypothesizing or conjecturing that the different inferential logics are implicated. After generating some the hypotheses, the scientist then tries to deduce their implications and design experiments. In the final step of the scientific process, he/she carries out the experiments to assess the viability and reliability of the hypotheses as probable explanations for the phenomena or phenomenal behavior. It is important to note, however, that no amount of testing can ever guarantee the truth of the probable explanation, but after surviving logical criticism and empirical testing, it could be accepted as a valid scientific theory. Continued cycles through this process over time lead to the development, acceptance, and/or rejection of scientific theories (Carnap 1953; Lakatos 1974).

3 Inferential Logics and the Generation of Hypotheses

The four inferential logics are useful for examining empirical observations of the phenomena or phenomenal behavior and logically reasoning about the observations to construct/invent hypotheses (plausible explanations) of various types which can be subjected to further examination and empirical testing. Some of the types of inferential logics, for example induction, have been criticized as an inadequate method for scientific discovery. However, more recent discussions view induction as part of process for constructing scientific theories in modern empirically based social science, in which no inferential logic is seen as an adequate method for scientific discovery. Inferential logics are simply means to making conjectures which will be subjected to logical criticism and empirical testing (Popper 2002). The four inferential logics illustrated below point to the different types of hypotheses which can assist the scientist as he or she tries to develop some theoretical explanation of the phenomena or phenomenal behavior of interest:

IL1. **Deduction** is the ***inference of a result*** from a rule and a case:

Rule—All men are mortal
Case—Socrates is a man
Result—Socrates is a mortal

IL2. **Induction** is the ***inference of a rule*** from the result and case:

Result—Socrates is a mortal
Case—Socrates is a man
Rule—All men are mortal

IL3. **Abduction** is the ***inference of the case*** from the rule and the result:

Rule—All men are mortal
Result—Socrates is a mortal
Case—Socrates is a man

IL4.**Retroduction** is the ***inference of the cause*** from the result and a case:

Result—Socrates is a mortal
Case—Socrates is dead
Cause—What killed Socrates

Inferential logics are approaches to "studying the facts and devising a theory to explain them" (Peirce, CP 5.145). For example, deduction can assist the social scientist in inferring (conjecturing) logical implications of social rules to design observation experiments which would test them. Induction can help the social scientist to infer general social rules from the observation of regularities in phenomena or phenomenal behavior. Abduction can help the social scientist categorize social phenomena or social behavior. And retroduction can help the social scientist infer causes or mechanisms underlying events or social behavior (as is the objective of the critical realist approach to social science). From the perspective of social scientific inquiry, the outcomes of any of these inferential logics can be seen as nothing more than hypotheses which will be empirically tested in pursuit of the development of a theory. What they do offer the social scientist is different ways of reasoning with initial empirical observations in the presence or absence of existing theories to propose new or alternative hypotheses for the development and testing of new theories (Fann 1970), which is fundamental to the growth of knowledge. The importance of generating hypotheses cannot be over stated; as Popper[1] observed, to systematically test any theory we must overcome the "numerical paucity of hypotheses" in a reasoned way. In Popper's view, science would stagnate and lose its empirical character if we failed (a) to obtain refutations of our theories and (b) to obtain verifications of new predictions.

4 Test Worthiness of Hypotheses

Is it worthwhile to test every hypothesis that is generated? Or should they be subjected to logical criticism even before we consider them for empirical testing? Both Peirce and Popper outlined some useful criteria for evaluating the test worthiness of hypotheses generated for the purpose of theory development. Peirce (CP, 1931–1958) suggests two criteria: (1) the likelihood that the hypothesis will be confirmed in testing, that is, a numeric estimate of the probability that the

[1] It is important to note here that Popper commonly uses the term statement or scientific statement to mean hypothesis. In "The Logic of Scientific Discovery" (LSD pp. 94), he states, "For we can utter no scientific statement that does not go far beyond what can be known with certainty 'on the basis of immediate experience'. … every statement has the character of a theory, of a hypothesis." In LSD pp. 32, he states, "For a new idea, put up tentatively, and not yet justified in any way—an anticipation, a hypothesis, a theoretical system, or what you will- conclusions are drawn by logical deduction."

hypothesis will explain the facts; (2) the breadth of the hypotheses; as he puts it, *twenty skillful hypotheses will ascertain what a million stupid ones will fail to do.* Popper (2002) on the other hand suggests three requirements for entertaining and exploring alternative hypotheses[2]: (1) The new hypotheses should proceed from some simple, new, and powerful idea about some connection or relationship between hitherto unconnected facts. (2) The new hypotheses should lead to prediction of the phenomena which have not so far been observed and should be independently testable. (3) The new hypotheses must pass some test that shows that it is likely to provide successful predictions (see also Popper 1957). For Popper, these three criteria are an important defense against the generating hypotheses, "which in spite of their increasing degree of testability, were ad hoc, and would not get us any nearer to the truth" (ibid, pp. 331).

These test worthiness criteria for hypotheses have been operationalized using probabilistic reasoning (Peirce 1931–1958; Popper 1957 AIM). The key issue, however, is to distinguish among the hypotheses the ones that offer potentially a better explanation of the facts. This approach is commonly called inference to the better explanation (IBE) (Harman 1965; Hempel 1965, Day and Kincaid 1994). The IBE rule suggests that H is preferred when from a set of hypotheses H is a better explanation of the evidence E relative to the background assumptions B. This rule can be defined as follows:

A5. Given the evidence E accept the hypothesis H which maximizes the likelihood P(E/H&B)

Since we are looking for testable hypotheses, we are not interested in the case in which the hypothesis H is a universal generalization such that $P(E/H\&B) = 1$. Such a hypothesis would be not be falsifiable and as such would be invalid for the purpose of scientific theory construction and testing. Further, because we are working with evidence collected using instruments based on existing theory, we are unlikely to see the situation in which we have an hypothesis H where $P(E/H\&B) = 0$. This would imply that the observational data E are all together incompatible with the hypothesis H. A solution to these two problems can be formally defined as follows:

A6. Given the evidence E accept the hypothesis H which maximizes P(H/E)—P(H).

By specifying this rule, we can ensure that the chosen hypothesis has a high posterior probability and high explanatory power, as we can now set some cutoff point for $P > 0$ that satisfies our interest for exploring the research. This brings us to the final theoretical problem of how to derive more general hypotheses from lower level ones. A problem commonly referred to as entailment (Hempel 1965; Hintikka 1968). Entailment deals with the basic problem of deriving more general hypotheses from lower level hypotheses. For both Peirce and Popper, if hypothesis H logically includes hypotheses H1 and H2, and H is falsifiable, then H is chosen

[2] Three Requirements for The Growth of Knowledge, in Popper, K. Conjectures and Refutations, 2002 Edition, pp. 326–327.

as the more general hypothesis. Niiniluoto and Tuomela (1973) have argued that this rule can be rewritten as the IBE rule: if hypothesis H explains the evidence E better that H1 and H2 combined, H is the more general hypotheses. Further, we can also satisfy the requirements for assessing the posterior probability of our hypotheses by defining the following constraint:

A7. Assuming that P(H) > 0 and P(E) < 1, if H entails E, then P(H/E) > P(H).

To summarize, the empirically based approach to social scientific inquiry holds that a scientific theory is a set of hypotheses tentatively proposed with the objective of explaining phenomena or phenomenal behavior. And according to Popper (1968, 2002), the scientist must take risks and continually posit alternative hypotheses and rigorously and ruthlessly falsify them. The scientist must reject ad hoc hypotheses that would support the theory in favor of systematically derived ones that offer the potential for bold new predictions. The four inferential logics are strategies for reasoning in different ways about initial empirical observations to generate hypotheses which can be later subjected to logical criticism and systematic testing empirically. The steps in this process are as follows: (1) generating alternative hypotheses; (2) the evaluation of the test worthiness of the hypotheses; (3) selecting an appropriate set of the hypotheses that present an alternative or improved model to explain the evidence; and (4) the testing of the alternative model. Many of the methods in this book were developed based on individual inferential logics. The discussion and demonstrations of use of each method clearly identify which inferential mechanism underlies it.

Acknowledgments Some of the material in this chapter previously appeared in: "Using Decision Tree Modelling to Support Peircian Abduction in IS Research: A Systematic Approach for Generating and Evaluating Hypotheses for Systematic Theory Development," *Information Systems Journal 21:5*, 407–440 (2011).

References

Berger PL, Luckmann T (1991) The social construction of reality: a treatise in the sociology of knowledge (No. 10). Penguin, UK

Carnap R (1953) Testability and Meaning. In: Feigel H, Broddeck M (eds) Readings in the Philosophy of Science. Appleton-Century-Crofts, New York, pp 47–92

Day T, Kincaid H (1994) Putting inference to the best explanation in its place. Synth Int J Epistemol Methodol Philos Sci 98:271–295

Fann KT (1970) Peirce's theory of abduction. Martinus Nijhoff, Amsterdam

Harman G (1965) Inference to the best explanation. Philos Rev 74:88–95

Hempel CG, Oppenheim P (1965) Studies in the logic of explanation. Philos Sci 15:135–175

Hempel CG (1965) Aspects of scientific explanation. The Free Press, New York

Hintikka J (1968) The varieties of information and scientific explanation. In: van Rootselaar B, Staal JF (eds) Logic Methodology and Philosophy of Science III. North Holland, Amsterdam, pp 151–171

Lakatos I (1974) Falsification and the methodology of scientific research programs. In: Lakatos I, Musgrave A (eds) Criticism and the growth of knowledge. Cambridge University Press, Cambridge, pp 91–195

Niiniluoto I (1997) Reference invariance and truthlikeness. Proc Philos Sci 64:546–554
Niiniluoto I, Tuomela R (1973) Theoretical concepts and hypothetico-inductive inference. Dordrecht: D. Reidel PublishingCompany.
Peirce CS (1931–1958) Collected papers of Charles Sanders Peirce, vols 1–8. In: Hartshorne C, Weiss P, Burks A (eds) Harvard University Press, Harvard
Popper KR (1957) The aim of science. Ratio 1(1):24–35
Popper KR (1968) The logic of scientific discovery. Harper Torch Books, New York
Popper KR (2002) Conjectures and refutations. Routledge Classics Edition, London
Popper KR (1972) Objective knowledge: an evolutionary approach. Oxford University Press, Oxford
Thagard R (1978) The best explanation: criteria for theory choice. J Philos 65:76–92
Toulmin SE, Barzun J (1981) Foresight and understanding: an enquiry into the aims of science. Greenwood Press, Westport
Winch P (1958/1990) The idea of a social science: and its relation to philosophy. Routledge, London

Chapter 3
Overview on Decision Tree Induction

Kweku-Muata Osei-Bryson

The chapter provides an overview of decision tree (DT) induction. Its main purpose is to introduce the reader to the major concepts underlying this data mining technique, particularly those that are relevant to the chapters that involve the use of this technique.

1 Introduction

A DT is an inverted tree structure representation of a given decision problem such that each non-leaf node is associated with one of the *Predictor Variables*, each branch from a non-leaf node is associated with a subset of the values of the corresponding predictor variable, and each leaf node is associated with a value of the *Target* (or dependent) variable. There are two main types of DTs: (1) *Classification Trees* (CT) and (2) *Regression Trees* (RT). For a CT, the target variable takes its values from a discrete domain, while for a RT, the target variable takes its values from a continuous domain (e.g., Osei-Bryson 2004; Ko and Osei-Bryson 2002; Torgo 1999; Kim and Koehler 1995; Breiman et al. 1984).

1.1 Classification Tree

A DT can be described as a model of a prediction problem in the form of interpretable and actionable rules (see Fig. 1). Associated with each leaf of the DT is an IF–THEN rule. For a given rule, the condition component (independent variable(s)

K.-M. Osei-Bryson (✉)
Department of Information Systems, Virginia Commonwealth University, 301W Main Street, Richmond, VA 23284, USA
e-mail: KMOsei@VCU.Edu

K.-M. Osei-Bryson and O. Ngwenyama (eds.), *Advances in Research Methods for Information Systems Research*, Integrated Series in Information Systems 34,
DOI: 10.1007/978-1-4614-9463-8_3,

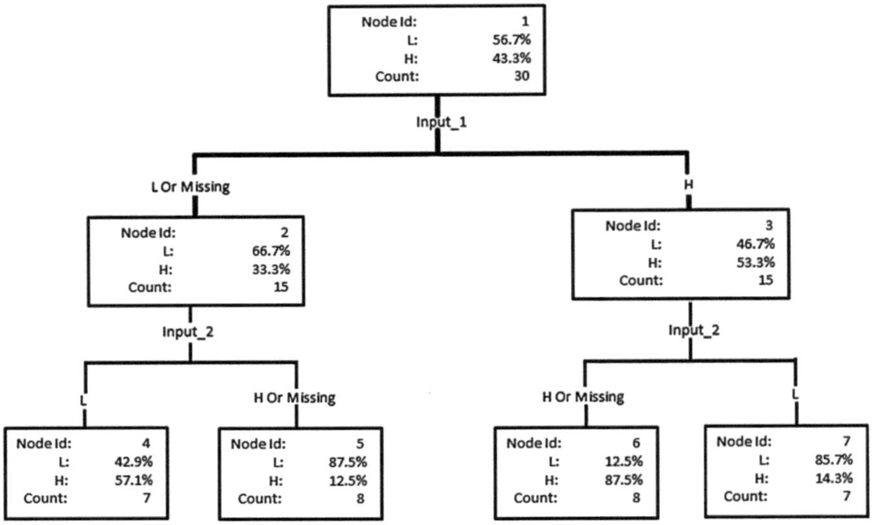

Fig. 1 Graphic of DT for illustrative example 1

and their values) of the rule is described by the set of relevant internal nodes and branches from the root to the given leaf; the action part of the rule is described by the relevant leaf, which provides the relative frequencies for each class of the dependent variable.

To generate a DT from a given dataset, a single variable must be identified as the *Target* (or dependent) variable and the potential predictors must be identified as the *Input* variables. We will illustrate with a play example that has 3 Input variables (Input_1, Input_2, Input_3) and a Target variable that has the name *Target*. The example dataset is described in Table 1. The SAS Enterprise Miner data mining software was applied to this dataset, resulting in the CT that is displayed in Fig. 1.

The reader may note that each branch of the DT is associated with a given variable (e.g., Input_1) and a value for that variable (e.g., Input_1 = "*L*"). Let us consider the leftmost leaf node, i.e., Node 4. The parent of this leaf node is Node 2. The root node of the DT is Node 1. From the root node to Node 4, there are two connecting branches associated with the simple condition Input_1 = "*L*" and the simple condition Input_2 = "*L*". The conjunction of these two simple conditions is the condition component of the rule that is associated with Node 4. For this rule, the *Target* class *H* is the predicted class for this rule because it has the largest relative frequency of the two classes (i.e., *L*: 42.9 %; *H*: 57.1 %) in the relevant leaf node, i.e., Node ID = 2.

Rules 1 and 2 are considered to be *sibling rules* because they are children of the internal node, Node_ID = 2, that is associated with the condition Input_1 = "*L*". Similarly, rules 2 and 3 are *sibling rules* because they are children of the internal node, Node ID = 3, that is associated with the condition Input_1 = "*H*" (Table 2).

Table 1 Dataset for illustrative example 1

Row ID	Input_1	Input_2	Input_3	Target	Row ID	Input_1	Input_2	Input_3	Target
1	*L*	*L*	*H*	*L*	16	*H*	*H*	*H*	*H*
2	*L*	*L*	*H*	*H*	17	*H*	*H*	*H*	*H*
3	*L*	*L*	*H*	*H*	18	*L*	*L*	*L*	*H*
4	*L*	*L*	*H*	*L*	19	*L*	*L*	*L*	*L*
5	*H*	*L*	*H*	*L*	20	*L*	*L*	*L*	*H*
6	*H*	*L*	*H*	*L*	21	*H*	*L*	*L*	*L*
7	*H*	*L*	*H*	*L*	22	*H*	*L*	*L*	*L*
8	*H*	*L*	*H*	*L*	23	*H*	*L*	*L*	*H*
9	*L*	*H*	*H*	*L*	24	*L*	*H*	*L*	*H*
10	*L*	*H*	*H*	*L*	25	*L*	*H*	*L*	*L*
11	*L*	*H*	*H*	*L*	26	*L*	*H*	*L*	*L*
12	*L*	*H*	*H*	*L*	27	*H*	*H*	*L*	*H*
13	*L*	*H*	*H*	*L*	28	*H*	*H*	*L*	*H*
14	*H*	*H*	*H*	*H*	29	*H*	*H*	*L*	*L*
15	*H*	*H*	*H*	*H*	30	*H*	*H*	*L*	*H*

Table 2 Rule set of CT of Fig. 1

Rule ID	Leaf node ID	Condition component of the rule	Target: distribution in leaf node		Target: predicted value
			L (%)	*H* (%)	
1	4	Input_1 = "*L*" and Input_2 = "*L*"	42.9	57.1	*H*
2	5	Input_1 = "*L*" and Input_2 = "*H*"	87.5	12.5	*L*
3	6	Input_1 = "*H*" and Input_2 = "*H*"	12.5	87.5	*H*
4	7	Input_1 = "*H*" and Input_2 = "*L*"	85.7	14.3	*L*

1.2 Regression Tree

A RT is a DT in which the target variable takes its values from a continuous domain (numeric). For each leaf, the RT associates the mean value and the standard deviation of the target variable. Similar to CTs, each RT has a corresponding rule set, an example of which is provided in Table 3.

- Rule 1 can be interpreted as follows: *IF LABEXP* < 6.204105 *THEN Average* (*V*) = 6.23593 and *SD* (*V*) = 0.6462, with 210 training set observations being associated with this rule.
- Rule 2 can be interpreted as follows: *IF LABEXP is Between* 6.204105 *and* 6.940255 *THEN the Average* (*V*) = 7.17834 and *SD* (*V*) = 0.38161, with 240 training set observations being associated with this rule.

Table 3 Rule set of a regression tree

Rule ID	Description
1	IF LABEXP < 6.204105 THEN V = {AVE: 6.23593; SD = 0.6462; COUNT = 210}
2	IF LABEXP ∈ [6.204105, 6.940255) THEN V = {AVE: 7.17834; SD = 0.38161; COUNT = 240}
3	IF LABEXP ∈ [6.940255, 8.009865) THEN V = {AVE: 8.01077; SD = 0.42665; COUNT = 272}
4	IF LABEXP ≥ 8.009865 THEN V = {AVE: 9.10683; SD = 0.61329; COUNT = 104}

AVE mean, *SD* standard deviation, *COUNT* number of observations

Although RTs are similar to regression, the RT model is more like a step function, whereas the regression model involves a continuous linear function. Compared to regression models, RTs provide a model with better interpretability because the model involves interpretable English rules or logic statements. There have been instances where a RT has shown clues to datasets, while a traditional linear regression analysis could not clearly indicate them (Breiman et al. 1984).

1.3 DT Generation Process

The DT generation process involves a *Growth Phase* and optionally a *Pruning Phase* (e.g., Kim and Koehler 1995). The *Growth Phase* involves generating a DT from the *Training* data such that either each leaf node is associated with a single class or further partitioning of the given leaf would result in the number of observations in one or both child nodes being below some specified threshold. The *Pruning Phase* aims to generalize the *Unpruned* DT that was generated in the *Growth Phase* in order to avoid over-fitting the final DT to the training data. Therefore, in this phase, the *Unpruned* DT is evaluated against what is referred to as *Validation* dataset in order to generate a subtree of the *Unpruned* DT generated in the *Growth Phase* that has the lowest error rate against the *Validation* data. It follows that this DT is not independent of the *Training dataset* or the *Validation data*. For this reason, it is important that the distribution of observations in the *Validation data* corresponds to the overall distribution of the observations.

To avoid over-fitting of the model to the training data, for large datasets the generation of a DT includes partitioning the model dataset into either two parts (i.e., *Training* and *Validation*) or three parts (i.e., *Training, Validation,* and *Test*). For small datasets, techniques such a v-fold (e.g., 10-fold) cross-validation allow for the entire dataset to be used for both the *Growth* and *Pruning* phases.

1.4 Recursive Splitting

The Growth Phase involves recursive splitting of the *Training* dataset into progressively smaller subsets that are more homogenous with respect to the Target variable. At each iteration, splitting decisions would be made automatically by a DT induction algorithm, which include

- On what variable to split
- What is the best split
- When to stop splitting

Recursive Splitting Example

We will use illustrative dataset of Sect. 1.1 to illustrate how recursive partitioning would be done. Table 4 displays the relevant splits and partitioning of the datase. The CT of Fig. 1 would have resulted from this process.

Table 4 Result of recursive splitting

Splits		Variables					
Split 1	Split 2	Row ID	Input_1	Input_2	Input_3	Target	Rule ID
Input_1 = "*L*"	Input_2 = "*L*"	2	*L*	*L*	*H*	*H*	1
		3	*L*	*L*	*H*	*H*	
		18	*L*	*L*	*L*	*H*	
		20	*L*	*L*	*L*	*H*	
		1	*L*	*L*	*H*	*L*	
		4	*L*	*L*	*H*	*L*	
		19	*L*	*L*	*L*	*L*	
	Input_2 = "*H*"	24	*L*	*H*	*L*	*H*	2
		9	*L*	*H*	*H*	*L*	
		10	*L*	*H*	*H*	*L*	
		11	*L*	*H*	*H*	*L*	
		12	*L*	*H*	*H*	*L*	
		13	*L*	*H*	*H*	*L*	
		25	*L*	*H*	*L*	*L*	
		26	*L*	*H*	*L*	*L*	
Input_1 = "*H*"	Input_2 = "*H*"	14	*H*	*H*	*H*	*H*	3
		15	*H*	*H*	*H*	*H*	
		16	*H*	*H*	*H*	*H*	
		17	*H*	*H*	*H*	*H*	
		27	*H*	*H*	*L*	*H*	
		28	*H*	*H*	*L*	*H*	
		30	*H*	*H*	*L*	*H*	
		29	*H*	*H*	*L*	*L*	
	Input_2 = "*L*"	23	*H*	*L*	*L*	*H*	4
		5	*H*	*L*	*H*	*L*	
		6	*H*	*L*	*H*	*L*	
		7	*H*	*L*	*H*	*L*	
		8	*H*	*L*	*H*	*L*	
		21	*H*	*L*	*L*	*L*	
		22	*H*	*L*	*L*	*L*	

1.5 Selection of the Splitting Method

A splitting method (e.g., Osei-Bryson and Giles 2004; Quinlan 1993; Taylor and Silverman 1993) is the component of the DT induction algorithm that determines both the attribute that is selected for a given node of the DT and also the partitioning of the values of the selected attribute into mutually exclusive subsets such that each subset uniquely applies to one of the branches that emanate from the given node. It is well known that there is no single splitting method that will give the best performance for all datasets.

While some datasets are insensitive to the choice of splitting methods, other datasets are very sensitive to the choice of splitting methods. Given that it is never known beforehand which splitting method will lead to the best DT for a given dataset, it is advisable that the data miner explores the effects of different splitting methods. For CTs, these include Chi, Gini, and various entropy-based methods. For RTs, these include variance reduction and F-test.

1.6 Prepruning and Post-pruning

1.6.1 Prepruning

Prepruning occurs during the Growth phase. Its goal is to ensure that the resulting DT is not over-fitted to the *Training* dataset. It attempts to stop the growth of the DT if the node is pure, or the number of examples is below some threshold, or the split is not statistically significant at a specified.

1.6.2 Post-pruning

Post-pruning (e.g., Osei-Bryson 2007; Fournier and Cremilleux 2002) occurs during the *Pruning Phase* of DT induction. A sequence of sub-trees of decreasing sizes and complexity are generated from the unpruned DT that was generated in the *Growth phase*; then an assessment criterion (e.g., best value of assessment measure such as error rate), specified number of leaves, maximum number of leaves) is applied to the Validation data to determine the "best" subtree. It follows that the selected subtree is not independent of the *Training* dataset or the *Validation* dataset. For this reason, it is important that the distribution of observations in the *Validation* dataset corresponds to the overall distribution of the observations.

2 Software Implementation of the DT Generation Process

Many DM software packages (e.g., C5.0, SAS Enterprise Miner, and IBM Intelligent Miner) provide facilities that make the generation of DTs a relatively easy task. Below we present a set of figures (i.e., Fig. 2a–c) that describe a sample DT

(a)

Label	Role	Description
Outcome	Target	Two Classes: *Bad*: Defaulted or seriously delinquent; *Good*
loan	Input	Amount of Current Loan Request
mortdue	Input	Amount Due on existing Mortgage
value	Input	Value of current property
reason	Input	Home improvement or debt consolidation
job	Input	Job
yoj	Input	Years on current job
derog	Input	Number of major derogatory reports
deling	Input	Number of delinquent trade lines
clage	Input	Age of oldest trade line
ninq	Input	Number of recent credit inquiries
clno	Input	Number of trade (credit) lines
debtinc	Input	Debt to Income Ratio

(b)

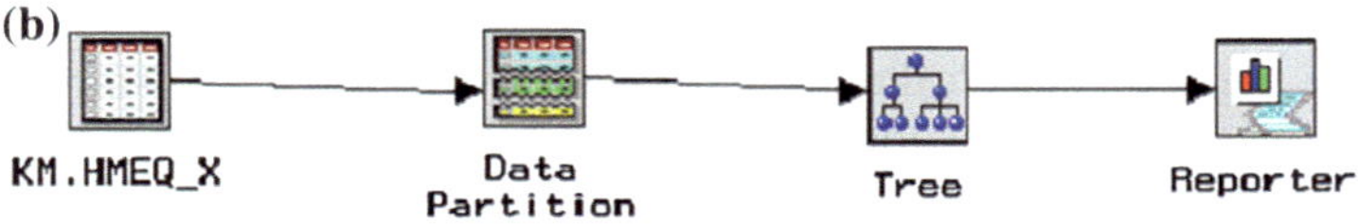

(c)

RuleID **Description**

1 IF Debt_to_Income_Ratio (i.e. debtinc) < 43.724
THEN Outcome ={GOOD: 93.3%; BAD: 6.7%}

2 IF Debt_to_Income_Ratio (i.e. debtinc) ≥ 45.723
THEN Outcome ={GOOD: 0.0%; BAD: 100.0%}

3 IF 0.5 ≤ Number_of_Delinquent_Trade_Lines (i.e. delinq) < 2.5
& 43.724 ≤ Debt_to_Income_Ratio (i.e. debtinc) < 45.723
THEN Outcome ={GOOD: 24.8%; BAD: 75.2%}

4 IF Number_of_Delinquent_Trade_Lines (i.e. delinq) ≥ 2.5
& 43.724 ≤ Debt_to_Income_Ratio (i.e. debtinc) < 45.723
THEN Outcome ={GOOD: 6.0%; BAD: 94.0%}

5 IF Age_of_Oldest_Trade_Lines_in_Months (i.e. clage) < 188.75
& Number_of_Delinquent_Trade_Lines (i.e. delinq) < 0.5
& 43.724 ≤ Debt_to_Income_Ratio (i.e. debtinc) < 45.723
THEN Outcome ={GOOD: 38.4%; BAD: 61.6%}

6 IF Number_of_Major_Derogatory_Reports (i.e. derog) < 0.5
& Age_of_Oldest_Trade_Lines_in_Months (i.e. clage) ≥ 188.75
& Number_of_Delinquent_Trade_Lines (i.e. delinq) < 0.5
& 43.724 ≤ Debt_to_Income_Ratio (i.e. debtinc) < 45.723
THEN Outcome ={GOOD: 78.8%; BAD: 21.2%}

7 IF 0.5 ≤ Number_of_Major_Derogatory_Reports (i.e. derog)
& 188.75 ≤ Age_of_Oldest_Trade_Lines_in_Months (i.e. clage)
& Number_of_Delinquent_Trade_Lines (i.e. delinq) < 0.5
& 43.724 ≤ Debt_to_Income_Ratio (i.e. debtinc) < 45.723
THEN Outcome ={GOOD: 47.4%; BAD: 52.6%}

Fig. 2 **a** Description of sample input dataset. **b** SAS Enterprise Miner process flow diagram for generating sample DT. **c** Set of rules associated with the DT

induction application using the SAS Enterprise Miner software. The given decision problem involves predicting whether a given loan applicant will default on the loan.

- Figure 2a describes the input data.
- Figure 2b presents the SAS Enterprise Miner *process flow diagram* that is used to generate the DT from the given dataset. Each node in this diagram performs a specific task: The input node reads the given dataset; the data partition node partitions the input dataset into *Training*, *Validation*, and *Test* datasets; the tree node generates the CT; and the reporter node prepares a report of the output including the set of rules that are associated with the CT.
- Figure 2c provides the set of IF–THEN rules that are associated with the CT. The predicted class for a given rule is the one with the largest posterior probability.
- Rule 1 can be interpreted as follows: *IF Debt_to_Income_Ratio* < 43.724 *THEN the Outcome is likely to be GOOD with relative frequency* 0.933 and *BAD with relative frequency* 0.067; the predicted class would be GOOD since it has the highest relative frequency for this rule.
- Rule 2 can be interpreted as follows: *IF Debt_to_Income_Ratio* $\geq$ 45.723 *THEN the Outcome is likely to be GOOD with relative frequency* 1.000 and *BAD with relative frequency* 0.000; the predicted class would be BAD since it has the highest relative frequency for this rule.

References

Breiman L, Friedman J, Olshen R, Stone J (1984) Classification and regression trees. Wadsworth Inc., Belmont

Fournier D, Cremilleux B (2002) A quality index for decision tree pruning. Knowl-Based Syst 15:37–43

Kim H, Koehler G (1995) Theory and practice of decision tree induction. Omega 23(6):637–652

Ko M, Osei-Bryson K-M (2002) A regression tree based exploration of the impact of information technology investments on firm level productivity. European conference of information systems, pp 507–517

Osei-Bryson K-M (2004) Evaluation of decision trees: a multi-criteria approach. Comput Oper Res 31(11):1933–1945

Osei-Bryson K-M, Kendall Giles K (2004) An exploration of a set entropy-based hybrid splitting methods for decision tree induction. J Database Manage 15(3):1–17

Osei-Bryson K-M (2007) Post-pruning in decision tree induction using multiple performance measures. Comput Oper Res 34(11):3331–3345

Quinlan J (1993) C4.5 Programs for machine learning. Morgan Kaufmann, San Mateo

Taylor P, Silverman B (1993) Block diagrams and splitting criteria for classification trees. Stat Comput 3(4):147–161

Torgo L (1999) Predicting the density of algae communities using local regression trees. Proceedings of the European congress on intelligent techniques and soft computing (EUFIT'99)

Chapter 4
An Approach for Using Data Mining to Support Theory Development

Kweku-Muata Osei-Bryson and Ojelanki Ngwenyama

The rapid and constant change in information technologies (IT), organizational forms, and social structures is challenging our existing theories of the impact IT on organizations and society. A basic problem for researchers is how to generate testable hypotheses about the given area of research. However, new IT offer opportunities for information processing and problem solving that could extend the capacity of researchers to generate hypotheses and systematically explore the limitations of any theory. The idea of using IT to support IS research is not new. In this chapter, we explore and illustrate how data mining techniques could be applied to assist researchers in systematic theory testing and development.

1 Introduction

A fundamental tenet of positivist scientific inquiry is the exploration of the limits of existing theories in order to postulate alternatives (Popper 1957). However, it has been observed that a significant obstacle to such exploration of theories is the difficulty that the scientist faces in continually generating hypotheses for (rigorous and ruthless) testing (Popper 1968). In any scientific discipline, testing the limits of existing theories is a difficult, time-consuming, and costly process. In information systems (IS), these difficulties are further confounded by rapidly changing information technologies and the emergent nature of the organizations that utilize them.

Currently, much of IS theorizing is built on theories from other disciplines and little systematic investigation (as defined by Popper) is conducted to explore the limits of

K.-M. Osei-Bryson (✉)
Department of Information Systems, Virginia Commonwealth University,
301 West Main Street, Richmond, VA 23284, USA
e-mail: KMOsei@VCU.Edu

O. Ngwenyama
Ted Rogers School of Management, Ryerson University, 350 Victoria Street, Toronto,
ON M5B 2K3, Canada
e-mail: Ojelanki@ryerson.ca

K.-M. Osei-Bryson and O. Ngwenyama (eds.), *Advances in Research Methods for Information Systems Research*, Integrated Series in Information Systems 34,
DOI: 10.1007/978-1-4614-9463-8_4,

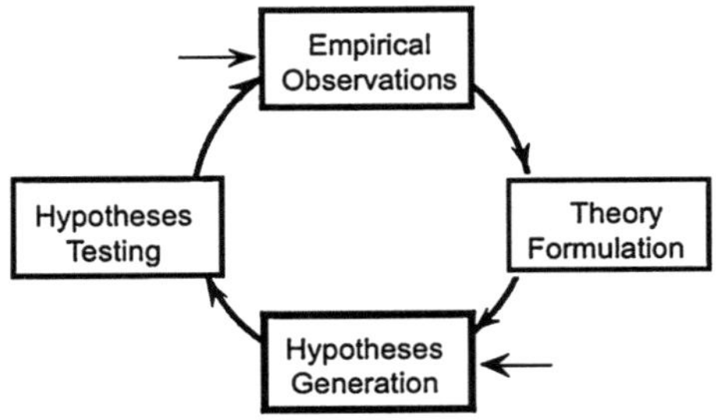

Fig. 1 General model of H-D logic (Chalmers 1994; Grimes 1990; Palys 2003)

those theories within the context of IS phenomena. According to Popper (1963), systematic testing should involve not only attempts to falsify a theory via repeated observation and experimentation, but to propose alternative hypotheses that would also be subject to falsification.[1] The second part of Popper's prescriptions, which is rigorous exploration of the limits of the theory, is much more difficult to attain due to limitations on the generation of alternative hypotheses. Such systematic testing of IS theories is even more challenging due to the dynamic nature of the phenomena (organizations and technology) about which the IS discipline is concerned. In this chapter, we discuss how data mining technologies could be applied to assist researchers in abducting and evaluating hypotheses for systematic theory testing and development. Using IT to support IS research is not new; IS researchers use various IT-based tools such as EQS, AMOS, PLS (and others) for factor analysis and testing of models, and Atlas/TI, HyperResearch, NVivo, and Leximancer for qualitative data analysis. We believe that the method and computational procedure that we are presenting could improve the efficiency and effectiveness of research, not only in IS but in other disciplines. To illustrate our approach, we apply it to the well-known research area of user performance.

2 Theoretical Foundations of the Method

The general model of hypothetico-deductive (H-D) theory development, on which positivist IS research is based, can be described as a cyclical process (cf. Fig. 1) of empirical observation, theory formulation, hypotheses generation, and hypotheses testing (Chalmers 1994; Grimes 1990; Palys 2003).

A key limitation of the H-D model is the reliance on human imagination for the hypotheses generation phase (Popper 1963; Peirce 1867; cf. CP). However, we will demonstrate that a data mining technique known as decision tree generation can be adapted to support Peirce's method of hypotheses generation and help overcome this limitation. Data mining technology has already had a profound impact on scientific research in medicine and genetics (Brusic and Zeleznikow 1999; Lee and Irizarry 2001). Some of these techniques can enable IS researchers to identify and understand key relationships among empirical observations that can otherwise elude them.

[1] Science: Conjectures and Refutations, in Popper, K. Conjectures and Refutations, 2002 Edition, pp. 70–71.

Our method focuses on the hypotheses generation phase in Fig. 1. What we are proposing is a data mining approach for the abduction and evaluation of hypotheses based on Peirce's scientific method and specifically his theory of abduction. Some scholars (Putnam 1982; Quine 1995; Dipert 1995; Tursman 1987) consider Peirce a pioneer of the application of deductive and abductive logics to modern scientific inquiry. Other scholars consider him the founding father of modern deductive logic and the first person to systematize abductive inference into a formal method (Hanson 1961; Harman 1965; Hintikka 1968, 1997; Fann 1970; Quine 1995; Niiniluoto 1993, 1999). A fundamental objective of Peirce's work was the development of a scientific method, "*laws of development of science .... that rests on a sound general theory of logic.*" And while his method is similar to the hypothetico-deductive approach, he makes explicit a method for hypotheses generation and evaluation (cf. CP, 1 pp. 492). Our approach is based on Peirce's method and is implemented as a cyclic three-step procedure:

1. Abduction of a set of alternative hypotheses;
2. The evaluation of the test worthiness of the hypotheses;
3. Selection of an appropriate set of the hypotheses that present an alternative or improved model to explain the evidence.

3 The Data Mining-Based Technique

We will now explain how a data mining technique known as decision tree (DT) generation (Kim and Koehler 1995; Quinlan 1986) can be adapted to implement the logical theory of hypotheses generation and evaluation outlined above. In our methodology, we employ DT generation as a data analysis technique to support

1. The abduction of hypotheses from the empirical observations;
2. Entailment, the abduction of more general hypotheses from lower-level ones;
3. Computing the posterior probabilities and other test statistics for evaluating the test worthiness of the generated hypotheses;
4. The generation of the theoretical model.

3.1 Basic Concepts of Decision Trees

A decision tree can be described as a tree-structure model of a prediction problem in the form of interpretable and actionable rules (see Fig. 2). Associated with each leaf of the decision tree is an IF-THEN rule. For a given rule, the condition component (independent variable(s) and their values) of the rule is described by the set of relevant internal nodes and branches from the root to the given leaf; the action part of the rule is described by the relevant leaf, which provides the relative frequencies for each class of the dependent variable.

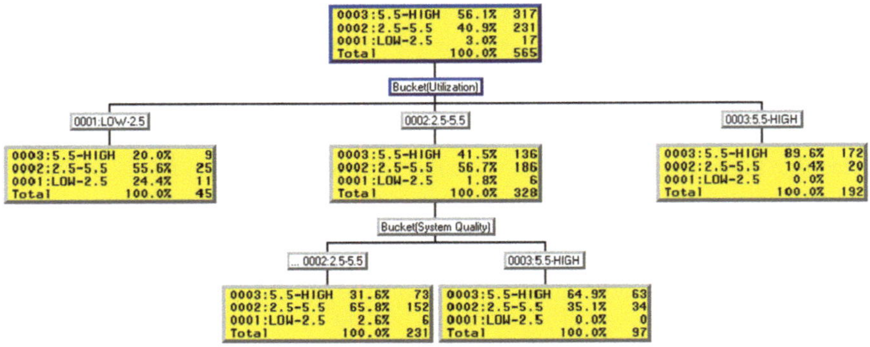

Fig. 2 Diagram of a decision tree

We display the four (4) rules for the decision tree in Fig. 2 where each variable has been discretized into three classes: Low: [1.0, 2.5]; Medium: [2.5, 5.5]; and High: [5.5–7.0]. It should be noted that *N* is the number of cases associated with the condition component of the rule.

These rules describe a set of hypotheses that were generated from our analysis; we will discuss them in detail later. Here, we limit our discussion to pointing out key characteristics relevant to understanding their use in our method. As stated earlier, these rules can be interpreted as conditional hypotheses. For example, Rule 1 can be interpreted as follows: *If the independent variable UTILIZATION is Low* (i.e., [0.0–2.5]) *then the dependent variable PERFORMANCE will be High* (i.e., [5.5–7.0]) *with relative frequency* $f = 0.200$, and 45 cases supporting this rule. Rule 4 can be interpreted as follows: IF *the independent variable UTILIZATION is High* (i.e., [5.5–7.0]) THEN *the dependent variable PERFORMANCE will be High* (i.e., [5.5–7.0]) *with relative frequency* $f = 0.896$, and 192 cases supporting this rule. On careful analysis, the reader will observe that Rules 2 and 3 have a similar structure, and they emanate directly from the internal node *SYSTEM QUALITY* that is itself connected to node *Utilization* through the branch *UTILIZATION* = Medium (i.e., [2.5–5.5]). Rules 2 and 3 are called ***sibling rules***, because the associated leaf nodes both have as parent the internal node *UTILIZATION* = Medium (i.e., [2.5–5.5]). We will return to our discussion of sibling rules, as they are important to the abduction of sibling rule hypotheses.

3.2 Abduction and Preliminary Evaluation of Hypotheses

We are now concerned with explaining how the theoretical concepts of abduction and evaluation are operationalized in the data mining-based method. The reader will recall that Peirce suggested that any abducted hypothesis should be evaluated for "the likelihood that the hypothesis will be confirmed in testing."

For the evaluation, we use traditional statistics-based hypothesis testing (illustrated later). We now focus on illustrating how we operationalize the process of abduction. We consider two types of hypotheses: (1) those derived using a single rule, which we will refer to as ***Single Rule Hypotheses*** and (2) those derived using a set of sibling rules, which we will refer to as ***Sibling Rules Hypotheses***. A Sibling Rules Hypothesis could be a *global hypothesis* or a *local hypothesis*.

3.2.1 Global and Local Hypotheses

A global hypothesis has the form: "*Variable **X*** has an *Impact_Type* impact on the *Target Variable **Y***," where Impact_Type ∈ {positive; negative; U-shaped curvilinear; inverted U-shaped curvilinear; none}. It reflects an average pattern that applies across the entire problem space. A local hypothesis has the form: "Given a certain *Backend Condition Event*, then *Frontend Condition Variable* ***X*** has an *Impact_Type* impact on the *Target Variable **Y***." An example of a local hypothesis is "Given *Utilization* = *Medium*, then *System_Quality* has a positive impact on *Performance*." It reflects an average pattern that applies across a subregion of the problem space. It is important to note that in some situations, while a *global hypothesis* might not be supported by empirical data, the *local hypothesis* might be. *Local hypotheses* that are supported by the data may be of interest to managers as they can provide guidance for action. Further, together with the associated global hypothesis, *local hypotheses* can provide a richer picture of the complex relationship between the given predictor variable and the target variable. This strategy provides researchers the ability to develop a detailed understanding of the dynamics of the problem domain that could lead to better (more relevant) predictive theories. It also operationalizes a key scientific principle, "inference to the better explanation." It is important to emphasize that it is unlikely that a researcher would be able to dream up all relevant *local hypotheses,* and consequently, without the strategy outlined above, relevant *local hypotheses* may not be conceptualized and interrogated in later confirmatory data analysis.

3.2.2 Discretization of Factor Scores

An important aspect of our systematic analysis of the hypotheses involves the use of discrete ordinal variables for the factor scores, originally coded on a 7-point Likert scale: [(1) strongly disagree; (2) moderately disagree; (3) slightly disagree; (4) neither agree nor disagree; (5) slightly agree; (6) moderately agree; and (7) strongly agree]. Discrete ordinal variables enable the researcher to derive meaningful ranges of values while abducting statistically testable hypotheses about the research problem. There are a variety of legitimate approaches to discretize the factor scores. One legitimate approach that is useful for our purposes, particularly with regard to the abduction of Sibling Rules Hypotheses, is discretization that involves 3 categories corresponding to *Low*, *Medium*, and *High* values for the given factor. Table 1 displays the qualitative and numeric intervals that we use for each category of the factor scores.

3.2.3 Sibling Rules Hypothesis

A *Sibling Rules Hypothesis* could be directional (e.g., variable *X* has a positive or negative impact on variable *Y*) or non-directional (i.e., variable *X* impacts variable *Y*). In either case, they are derived using a set of sibling rules (see Appendix B for more details). In our methodology, for any set of sibling rules, we can generate and test a corresponding *Sibling Rules Hypothesis*. For example, consider the Rules 2 and 3 (recall Table 2), which constitute a full set of sibling rules. Given this pair of rules, we could generate and indirectly test the directional *Sibling Rules Hypothesis*: "Given *Utilization* = *Medium*, then *System_Quality* has a positive impact on *Performance*." We could indirectly explore the validity of this Sibling Rules Hypothesis by testing the surrogate hypothesis: Given *Utilization* = *Medium*, target event *Performance* = *High* (i.e., [5.5–7.0]) is more likely to occur if *System_Quality* = *High* (i.e., [5.5–7.0]) than if *System_Quality* = *Low_Medium* (i.e., [1.0–5.5]), i.e., $p_H > p_{L\_M}$ where p_H, p_{L_M} are the population probabilities associated with the target event *PERFORMANCE* = *High* for *System_Quality* = *High* and *System_Quality* = *Low_Medium,* respectively. Acceptance (i.e., non-rejection) of this surrogate hypothesis would suggest that the given candidate *Sibling Rules* hypothesis might be valid and so should be abducted (see Table 3).

Table 1 Illustrative discretization intervals

Category	Intervals: regular coding		Intervals: reverse coding	
	Qualitative	Numeric	Qualitative	Numeric
Low (L)	Strongly disagree Moderately disagree	[1.0, 2.5]	Moderately agree Strongly agree	[5.5, 7.0]
Medium (M)	Slightly disagree Neither disagree nor agree Slightly agree	[2.5, 5.5]	Slightly disagree Neither disagree nor agree Slightly agree	[2.5, 5.5]
High (H)	Moderately agree Strongly agree	[5.5, 7.0]	Strongly disagree Moderately disagree	[1.0, 2.5]

Table 2 Rule set of decision tree of Fig. 2

Rule ID	Condition	Action: likely resulting *performance*	*N*
1	Utilization = Low	{*High* 20.0 %; *Med* 55.6 %; *Low* 24.4 %}	45
2	Utilization = Medium & System_Quality = Low_to_Med	{*High* 31.6 %; *Med* 65.8 %; *Low* 2.6 %}	231
3	Utilization = Medium & System_Quality = High	{*High* 64.9 %; *Med* 35.1 %; *Low* 0.0 %}	97
4	Utilization = High	{*High* 89.6 %; *Med* 10.4 %; *Low* 0.0 %}	192

Table 3 Candidate sibling rules hypotheses for DT of Fig. 2

ID	Condition events		*N*	Relative frequency (*f*)	Surrogate hypotheses			Abduct?	Candidate sibling hypothesis
	Backend	Frontend			Hypotheses	t-stat	Dec		
1		Util = L	45	0.200	$p_M > p_L$	2.77	A	YES	Util has a positive impact on PERFORMANCE
		Util = M	328	0.415	$p_H > p_M$	10.77	A		
		Util = H	192	0.896					
	○ Given the absence of a backend condition event, then the candidate Sibling Rules Hypothesis should be *global* ○ Given the directions of the relationships between the relative frequencies (i.e., $f_H > f_M > f_L$) of the frontend condition events, the Sibling Rules Hypothesis should involve a positive impact between the frontend variable [i.e., *Utilization* (*Util*)] and the target variable (i.e., *Performance*) ○ Each of the relevant surrogate hypotheses (i.e., $p_M > p_L$, $p_H > p_M$) has been accepted, so there is good reason to believe that the candidate sibling rules hypothesis should be accepted								
2	Util = M	SysQl = L_M	231	0.316	$p_H > p_{L\_M}$	5.59	A	YES	Given Util = M THEN SysQl has a positive impact on PERFORMANCE
		SysQl = H	97	0.649					
	○ Given the presence of a backend condition event, then the candidate Sibling Rules Hypothesis should be *local* ○ Given the direction of the relationship between the relative frequencies (i.e., $f_H > f_{L\_M}$) of the frontend condition events, the Sibling Rules Hypothesis should involve a positive impact between the frontend variable [i.e., *System_Quality* (*SysQl*)] and the target variable (i.e., *Performance*) ○ The relevant surrogate hypothesis (i.e., $p_H > p_{L\_M}$) has been accepted, so there is good reason to believe that the candidate Sibling Rules Hypothesis should be accepted.								

H: High; M: Medium; L_M: Low to Medium; L: Low
"Dec" (Decision): A = Accept; R = Reject

3.2.4 Single Rules Hypothesis:

A ***Single Rule Hypothesis*** would have the form: If *Condition* (e.g., *Utilization* is High) applies, then the *probability* of *target event* (e.g., User Performance is High) is strong (i.e., $p_0 \geq \tau_0$). It is important to note here that a *Condition* could consist of a single condition event (e.g., *Utilization* = *High*) or a conjunction of condition events (e.g., *Utilization* = *Medium* & *System_Quality* = *High*) in the given rule. For our purposes, we are only interested in *Single Rule* Hypotheses for which the value of p_0 satisfies the test worthiness specification of the researcher (i.e., $p_0 \geq \tau_0$). Therefore, for each set of sibling rules, a given rule is considered to be a strong rule only if statistical testing supports the hypothesis: $p_0 \geq \tau_0$. Only strong rules are used to generate Single Rule Hypotheses; so the first step involves identifying the strong rules in a given set of sibling rules. However, it is possible that each rule in a set of siblings could be strong, which would suggest that the discriminating variable for the given set of sibling rules would not be a useful predictor within the context of a Strong Single Rule hypothesis. Therefore, we only use those strong rules that are statistically different from at least one of their sibling rules to form Strong Single Rule Hypotheses.

3.3 *The Entailment Procedure*

The reader will recall that entailment is the abduction of more general hypotheses from lower-level ones. The basic logic of entailment outlined by Peirce and Popper posits: *If hypothesis H logically includes hypotheses H1 and H2, and H is falsifiable, then H is chosen as the more general hypothesis*. Our data mining-based method implements entailment via a strategy of merging child nodes of an internal node to form a new subtree. For example, if the nodes associated with rules 2 and 3 were merged, then we would obtain a new subtree (see Fig. 3 and Table 2) with the more general rule:

Rule 2_3 : IF *Utilization* is Medium THEN *Performance* = {High: 41.5 %; Medium: 56.7 %; Low: 1.8 %, where $N = 328$.

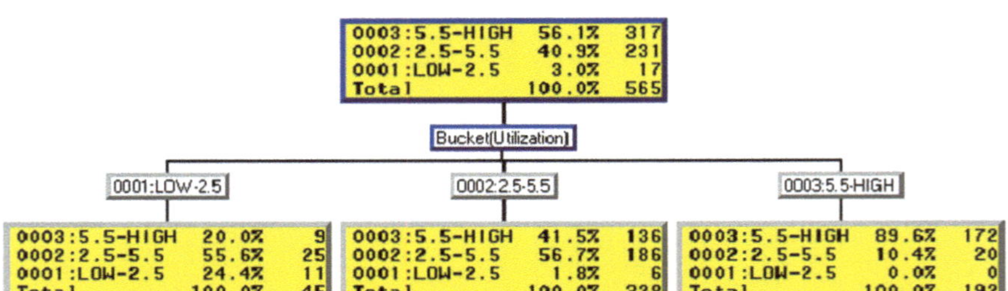

Fig. 3 Diagram of DT with Rules 2 and 3 combined

The reader may observe that this DT involves a partitioning of the dataset based on the *Utilization* variable, but in which the partitioning is determined by the DT generation procedure. The set of sibling rules that is associated with this DT allows us to explore the impact of *Utilization* on our performance variable *Performance*. However, every general rule entailed from sibling rules must be tested to ascertain whether it satisfies the test worthiness criterion. It is important to note here that commercial DM software such as SAS Enterprise Miner routinely provides the class distribution for each DT node, which can also be used for abducting higher-level hypotheses (entailment).

3.4 The Procedure for Abducting a Theoretical Model

We will now present our data mining methodology as a five-step procedure for abducting a theoretical model, in which Step 1 is done by the researchers and Steps 2–5 are automatic. We will first outline the procedure and then provide an illustrative example of the process (detailed step-by-step instructions for replication are provided in Appendix A). **Step 1** of the process is the preparation phase (activities "a" to "e" in Appendix A). In this phase, the researcher (a) identifies the dependent variables (or target variables in the language of data mining); (b) identifies any possible mediator variables; (c) defines and applies the discretization method such as that described in Table 1; (d) specifies data mining parameter values (e.g., splitting methods and minimum observations per leaf); (e) specifies the test worthiness threshold, τ_0, for Single Rule Hypotheses to be generated, and α, the significance level for statistical testing of the hypotheses; and (f) the researcher specifies the *Set of Decision Rules for Formulating and Evaluating Candidate Sibling Rules Hypotheses* (for details see Appendix B).

In Step 2 of the process, the data mining software then generates the decision tree rules using the data mining parameters defined by the researcher in Step 1. From the set of generated DT rules, all the sets of sibling rules are automatically identified for further analysis. For each set of sibling rules discovered for which the relationships between the relevant frequencies are included in the *Set of Decision Rules for Formulating and Evaluating Candidate Sibling Rules Hypothesis* (see Appendix B), a candidate Sibling Rules Hypothesis is automatically selected and indirectly tested using statistical difference of proportion tests to determine whether it is likely to be supported by the data. The candidate Sibling Rules hypotheses that satisfy the relevant difference in proportion tests are automatically abducted. In the second part of Step 2, Single Rules Hypotheses are automatically abducted and each is indirectly tested to determine whether it is likely to be supported by the data.

In **Step 3,** the software examines the set of abducted hypotheses to identify those potential mediator variables that are included in at least one abducted hypothesis for the dependent variable. For each such mediator variable, **Step 4**, which is analogous to **Step 2** for the dependent variables, is executed. Finally, in

Step 5, the theoretical model is generated by integrating the set of causal links between the predictor and dependent variables that were identified in the sets of abducted Sibling Rules Hypotheses and Single Rule Hypotheses. On the completion of this process, the researcher(s) should provide justification for the existence of each causal link of the integrated theoretical model, and links that they cannot justify should be removed.

4 Background on the Illustrative Problem

A number of studies have focused on end-user performance when using IS; in this section, we review some of this work that is relevant to our illustrative example. The body of research upon which we draw to illustrate our hypothesis generation methodology can be divided into two types of inquiry: (a) task–technology fit studies and (b) user satisfaction studies, each category approaching the study of end-user performance from a different perspective. Taken together, these studies have led to a list of factors and various models for investigating end-user performance and an inventory of instruments for eliciting data from end users. These studies also provide a list of variables/factors that have been investigated.

4.1 Data Collection and Factor Analysis

The data used to illustrate our methodology were collected from 20 organizations in two countries using well-known instruments that were validated by previous research. This involved the integration and use of three previously proposed questionnaires: (1) Goodhue and Thompson's (1995) task–technology fit instrument; (2) Etezadi-Amoli and Farhoomand's (1996) end-user computing satisfaction instrument (EUCS); and (3) Doll and Torkzadeh's (1988) EUCS instrument. In integrating, the three questionnaires overlapping items were removed, and new variables were added to capture geographic, industry, and demographic characteristics. The resulting final instrument[2] was validated. Factor analysis was conducted, and following this factor, scores were also generated[2]. Using the discretization transformations described in Table 1, for each observation, each corresponding factor score was then categorized as *High, Medium*, or *Low*.

[2] A discussion of the validation of the new instrument is beyond the scope of this paper. However, in validating the instrument, we used principal axis factoring and varimax rotation with Kaiser normalization. The reliability (alphas) for the groups of items were: System Quality, 0.9603; Documentation, 0.9337; Ease of Use 0.9266; Performance, 0.927; Utilization, 0.438; System Reliability, 0.8860; and Authorization 0.7618.

4.2 Application of Abduction Procedure to the Illustrative Problem

We used the SAS data mining software, Enterprise Miner, to generate and analyze the DTs, with the minimum number of cases (observations) for each DT rule set to 30. We also set the significance level for statistical testing at $\alpha = 0.05$ and set the threshold of our test worthiness for the Single Rule Hypotheses as $\tau_0 = 0.80$. The latter threshold implies that we only consider a rule to be strong if $p_0 > \tau_0 = 0.80$ and will only use these "strong" rules to form Single Rule Hypotheses. However, it should be noted that other rules may be generated during DT generation, but we will not use them to form Single Rule Hypotheses. The *Set of Decision Rules for Formulating and Evaluating Candidate Sibling Rules Hypothesis* is provided in Appendix B.

4.2.1 Abducting Hypotheses for the Dependent Variable

For our dependent variable, *Performance*, the selected target event is *High Performance* (H). We generated multiple DTs by means of a strategy that ensured that each variable was selected as the first split for at least one of these DTs, while having the lower-level splits to be automatically determined by the splitting algorithm. For each potential predictor variable *X* (i.e., *System Quality*, *Ease of Use, Reliability, Authorization, Documentation, Utilization*), this allowed us to indirectly test the global hypothesis:

Variable *X* has an *Impact_Type* impact on *Performance*

where *Impact_Type* $\in$ {No; positive; negative; U-shaped symmetric curvilinear; inverted U-shaped symmetric curvilinear}.

For each generated DT, the lower-level sets of sibling rules allowed for the selection and testing of candidate Sibling Rules Hypotheses. A complete description and analysis of the rule sets of all the generated DTs would be tiring for the reader. Therefore, we will limit our analysis and discussion to the rule sets of decision tree **A1E** (Table 4).

The candidate Sibling Rules Hypotheses for **A1E** were automatically selected based on the *Set of Decision Rules for Formulating and Evaluating Candidate Sibling Rules Hypotheses* in Appendix B. Each hypothesis was then indirectly tested for validity using statistical difference in proportions tests. Only those candidate Sibling Rules Hypotheses that passed the validity tests were abducted (see Table 4).

The next step in our procedure is the generation of the set of Single Rule Hypotheses for the dependent variable. For each rule in a set of sibling rules that is statistically different from each of its siblings at significance level $\alpha = 0.05$, we test the surrogate hypothesis $p_0 \geq \tau_0$ at significance level of $\alpha = 0.05$, where $\tau_0 = 0.80$. A Single Rule Hypothesis is abducted only if the corresponding surrogate hypothesis is accepted (i.e., not rejected). In Table 5, we display all the accepted Strong Single Rule Hypotheses and some of the rejected Single Rule Hypotheses. It should be noted that the target event is *High Performance*.

Table 4 The A1E rule set: Sibling Rules Hypotheses

Set ID	Condition event		N	Relative frequency (f)	Surrogate hypotheses			Abduct?	Candidate sibling rules hypothesis
	Backend	Frontend			Hypotheses	t-stat	Dec		
1		Util = L	45	0.200	$p_M > p_L$	2.77	A	YES	Util has a positive impact on PERFORMANCE
		Util = M	328	0.415	$p_H > p_M$	10.77	A		
		Util = H	192	0.896	$p_H > p_L$	9.89	A		
2	Util = H	SysQl = L_M	59	0.678	$p_H > p_{L\_M}$	6.56	A	YES	GIVEN Util = H THEN SysQl has a positive impact on PERFORMANCE
		SysQl = H	133	0.992					
3	Util = H & SysQl = H	EOU = L_M	47	0.979	$p_H > p_{L\_M}$	1.34	R	NO	GIVEN Util = H & SysQl = H THEN EOU has a positive impact on PERFORMANCE
		EOU = H	85	1.000					
4	Util = M	SysQl = L_M	231	0.316	$p_H > p_{L\_M}$	5.59	A	YES	GIVEN Util = M THEN SysQl has a positive impact on PERFORMANCE
	Util = M	SysQl = H	97	0.649					
5	Util = M & SysQl = L_M	Auth = L	44	0.394	$p_L > p_M$	1.80	A	YES	GIVEN Util = M & SysQl = L_M THEN Auth has a U-shaped curvilinear impact on PERFORMANCE
		Auth = M	157	0.255	$p_H > p_M$	2.89	A		

"Dec" (Decision): A = Accept; R = Reject

4.2.2 Abducting Hypotheses for the Mediator Variable

The first step in this phase of the process is to identify mediator variables. This involves determining whether the potential mediator variable (i.e., *Utilization*) is included in any of the abducted hypotheses for the dependent variable (i.e., *Performance*). For example, if we examine the abducted Sibling Rules Hypotheses (see Table 4) that were derived using *A1E*, we observe that *Utilization* (Util) is included in every abducted hypothesis. This suggests that *Utilization* is a mediator variable. Our next step is therefore to generate DTs that can be used to derive candidate hypotheses that involve *Utilization* as the target variable. Similar to the approach used for the dependent variable, candidate Sibling Rule Hypotheses were automatically selected and indirectly tested for the mediator variable. We present the results for one of the DTs in Table 6.

Similar to the approach used for the dependent variable, candidate Single Rule Hypotheses for the mediator variable were automatically selected and indirectly tested. For *Utilization*, all of the corresponding surrogate hypotheses were rejected and so no corresponding Strong Single Rule Hypothesis was generated (see Table 7).

4.3 Abduction of Theoretical Model

At this stage, a theoretical model can be automatically generated by integrating the set of causal links that are associated with the abducted Sibling Rules Hypotheses and Strong Single Rule Hypotheses. This theoretical model describes

Table 5 Abducted Single Rule Hypotheses for PERFORMANCE

DT	Condition event	N	f	t-stat	$p > 0.80$
A1E	Util = H	192	0.896	3.317	Accept
A1E	Util = H & SysQl = H	139	0.992	5.639	Accept
A1E	Util = H & SysQl = H & EOU = L_M	47	0.979	3.035	Accept
A1E	Util = H & SysQl = H & EOU = H	85	1.000	4.583	Accept
B1G	SysQl = H	236	0.835	1.341	Reject
B1G	SysQl = H & Auth = H	98	0.890	2.216	Accept
B1G	SysQl = H & Doc = H	77	0.900	2.179	Accept
B1E	EOU = H	158	0.842	1.316	Reject
B1E	EOU = H & SysRl = H	77	0.900	2.179	Accept
B1E	EOU = H & SysQl = H & SysRl = H	62	0.935	2.636	Accept
T1G1	Auth = H & SysQl = H	99	0.890	2.227	Accept
T1G1	Auth = M & SysQl = H & EOU = H	51	0.880	1.414	Reject
T1G2	Doc = H	103	0.835	0.884	Reject
T1G2	Doc = H & Auth = H	38	0.970	2.585	Accept
T1G3	SysRl = H & SysQl = H	13	0.880	0.693	Reject

Table 6 DT A2E: Sibling Rules Hypotheses

Set ID	Condition event		N	Relative frequency (f)	Surrogate hypotheses			Abduct?	Candidate sibling rules hypothesis
	Backend	Frontend			Hypotheses	t-stat	Dec		
1		EOU = L	48	0.063	$p_M > p_L$	2.90	A	YES	EOU has a positive impact on Utilization
		EOU = M	360	0.250	$p_H > p_M$	8.13	A		
		EOU = H	157	0.624	$p_H > p_L$	6.80	A		
2	EOU = H	SysQl = L_M	35	0.400	$p_H > p_{L\_M}$	3.11	A	YES	GIVEN EOU = H THEN SysQl has a positive impact on Utilization
		SysQl = H	122	0.689					
3	EOU = M	SysQl = L_M	250	0.176	$p_H > p_{L\_M}$	4.88	A	YES	GIVEN EOU = M THEN SysQl has a positive impact on Utilization
		SysQl = H	110	0.418					
4	EOU = M & SysQl = H	Auth = L_M	75	0.395	$p_H > p_{L\_M}$	0.75	R	NO	GIVEN EOU = M & SysQl = H THEN Auth has a positive impact on Utilization
		Auth = H	34	0.471					

Dec (Decision): A = Accept; R = Reject

Table 7 Sample of tested Single Rule Hypotheses for utilization

DT	Condition event	Target event	N	f	t-stat	$p > 0.80$
A2E	EOU = H	UTIL = H	157	0.624	−5.4956	Reject
A2E	EOU = H & SysQl = H	UTIL = H	122	0.689	−3.0525	Reject
T2G4	SysQl = H	UTIL = H	236	0.555	−9.3894	Reject
T2G4	SysQl = H & Auth = H	UTIL = H	98	0.663	−3.3732	Reject

the independent, mediator, and dependent variables and the newly hypothesized relationships. These hypotheses can be empirically tested in future investigations. We display our new theoretical model with causal links and associated supporting hypotheses in Table 8.

Table 8 Summary of abducted hypotheses

Causal link	Supporting abducted hypotheses
Auth ⇒ Perf	**Auth** has a symmetric U-shaped curvilinear impact on PERFORMANCE
	GIVEN Util = M & SysQl = L_M THEN **Auth** has a symmetric U-shaped curvilinear impact on PERFORMANCE
	GIVEN SysQl = L_M & EOU = M THEN **Auth** has a positive impact on PERFORMANCE
	GIVEN Doc = M THEN ***Auth*** has a symmetric U-shaped curvilinear impact on PERFORMANCE
	GIVEN Doc = H & **Auth = H** THEN PERFORMANCE = H with High Probability
	GIVEN SysQl = H & ***Auth = H*** THEN PERFORMANCE = H with High Probability
	GIVEN **Auth = H** & SysQl = H THEN PERFORMANCE = H with High Probability
Doc ⇒ Perf	**Doc** has a positive impact on PERFORMANCE
	GIVEN SysQl = H & **Doc = H** THEN PERFORMANCE = H with High Probability
	GIVEN Doc = H & Auth = H THEN PERFORMANCE = H with High Probability
EOU ⇒ Perf	**EOU** has a positive impact on PERFORMANCE
	GIVEN SysRl = H & SysQl = H THEN **EOU** has a positive impact on PERFORMANCE
	GIVEN **EOU = H** & SysRl = H THEN PERFORMANCE = H with High Probability
	GIVEN **EOU = H** & SysQl = H & SysRl = H THEN PERFORMANCE = H with High Probability
SysQl ⇒ Perf	**SysQl** has a positive impact on PERFORMANCE
	GIVEN Util = H THEN **SysQl** has a positive impact on PERFORMANCE
	GIVEN Util = M THEN **SysQl** has a positive impact on PERFORMANCE
	GIVEN EOU = M THEN **SysQl** has a positive impact on PERFORMANCE
	GIVEN Auth = M THEN **SysQl** has a positive impact on PERFORMANCE
	GIVEN Auth = H THEN **SysQl** has a positive impact on PERFORMANCE
	GIVEN SysRl = H THEN **SysQl** has a positive impact on PERFORMANCE
	GIVEN SysRl = L_M THEN **SysQl** has a positive impact on PERFORMANCE
	GIVEN **SysQl = H** & Auth = H THEN PERFORMANCE = H with High Probability
	GIVEN **SysQl = H** & Doc = H THEN PERFORMANCE = H with High Probability
	GIVEN **EOU** = H & *SysQl = H* & SysRl = H THEN PERFORMANCE = H with High Probability

(continued)

Table 8 (continued)

Causal link	Supporting abducted hypotheses
SysRl ⇒ Perf	**SysRl** has a positive impact on PERFORMANCE
	GIVEN EOU = H & **SysRl = H** THEN PERFORMANCE = H with High Probability
	GIVEN EOU = H & SysQl = H & **SysRl = H** THEN PERFORMANCE = H with High Probability
Util ⇒ Perf	**Util** has a positive impact on PERFORMANCE
	GIVEN **Util = H** THEN PERFORMANCE = H with High Probability
	GIVEN **Util = H** & SysQl = H THEN PERFORMANCE = H with High Probability
	GIVEN **Util = H** & SysQl = H & EOU = L_M THEN PERFORMANCE = H with High Probability
	GIVEN **Util = H** & SysQl = H & EOU = H THEN PERFORMANCE = H with High Probability
Auth ⇒ Util	GIVEN Doc = H THEN **Auth** has a positive impact on Utilization
Doc ⇒ Util	**Doc** has a positive impact on Utilization
	GIVEN Auth = H & EOU = H THEN **Doc** has NO impact on Utilization
EOU ⇒ Util	**EOU** has a positive impact on Utilization
	GIVEN SysQl = H THEN **EOU** has a positive impact on Utilization
	GIVEN SysQl = L_M THEN **EOU** has a positive impact on Utilization
	GIVEN Auth = H THEN **EOU** has a positive impact on Utilization
	GIVEN Doc = M THEN **EOU** has a positive impact on Utilization
	GIVEN SysRl = H THEN **EOU** has a positive impact on Utilization
SysQl ⇒ Util	GIVEN EOU = H THEN **SysQl** has a positive impact on Utilization
	GIVEN EOU = M THEN **SysQl** has a positive impact on Utilization
	GIVEN SysRl = L THEN **SysQl** has a positive impact on Utilization
	GIVEN SysRl = M THEN **SysQl** has a positive impact on Utilization
	GIVEN Auth = M THEN **SysQl** has a positive impact on Utilization
	SysQl has a positive impact on Utilization

High probability: $p > \tau = 0.80$

5 Conclusion

In this chapter, we presented an approach to systematic theory development and testing based on Peirce's scientific method. We will now consider some questions that might concern the reader about our DT-based approach:

1. Is this DT-based approach defensible from a statistical analysis perspective?

Our DT-based approach generates two types of hypotheses, Single Rule Hypotheses and Sibling Rules Hypotheses, and explanatory models. Hypotheses of these types can be subjected to traditional statistical hypothesis testing procedures. The main difference between these types of hypotheses and those that are typically used in IS studies is that they are derived from and grounded in empirical data, while the traditional ones are typically derived from existing theory and/or the researcher's imagination.

2. Are the hypotheses generated any good and could they be generated by the researcher without this method?

Several of the hypotheses generated by our approach meet the standards set out by Peirce: (1) They identify some connection or relationship in the data that were previously unidentified or overlooked. (2) They lead to predictions of phenomena, which until now have not been theorized about and independently investigated. (3) They have been evaluated to ensure their test worthiness. They all satisfy the Peircean posterior probability test that demonstrates that they are likely to be confirmed in empirical testing. The same cannot be said of hypotheses imagined by the researcher. While some of the hypotheses generated by our DT method might easily be imagined by the researcher, there are several others that are unlikely for a researcher to imagine. For example, hypotheses such as: "*Authorization* (Auth) *has a U-shaped curvilinear impact on Performance,*" and "*Given Authorization = High and Ease_of_Use = High then Documentation has NO impact on Utilization.*" On close examination, the reader will notice that these two hypotheses are relevant to theorizing about the phenomena and can offer advice to managers in organizational situations.

3. Can the average IS researcher use this method and will it make scientific work more productive?

Our method is general and applicable to any type of positivist theory development in IS research. It can be implemented using commercial data mining software packages (e.g., C5.0, SAS Enterprise Miner), which provide facilities for the generation of DTs, a relatively easy task. Given this fact, a major decision to be made by the researcher is the determination of target events (e.g., *Performance* is in the [5.5–7.0] interval) that are of interest. Once this decision has been made, many DM software software packages provide convenient facilities for discretizing the variables. Alternately, the discretization could be done using readily available data analysis tools such as in Excel. Further, all of the statistical testing for evaluating the generated hypotheses can be done automatically within the data mining software, if the researcher specifies the necessary tests and appropriate parameters.

In addition to generating meaningful hypotheses that are *likely to be valid* in future empirical testing, another important contribution of our approach is the ability to understand the functional form of causal relationships. This understanding would assist the researcher in making appropriate choices of statistical methods for testing hypotheses. For example, if the researcher comes to understand from using our method that certain causal relationships are nonlinear, then he or she would know that linear data analytic methods (such as partial least square approaches) are not adequate for interrogating them. It has already been shown that when linearity is incorrectly assumed for theoretical model relationships, empirical findings are often contradictory, confounding, or incorrect (Ko and Osei-Bryson 2004a, b). But the general practice of most positivist IS research is to assume that all relationships in theoretical models are linear and unconditional. Our approach could assist the researcher in discovering the correct functional form of the relationships and as a consequence make appropriate choices of statistical methods for testing the hypothetical models. Finally, our approach could be used in a complementary manner with the current dominant approach to confirmatory analysis. For example, given a dataset that has been previously used by a researcher for confirmatory data analysis, our DT-based approach could be used to abduct new hypotheses that would be used in future research using new datasets. In this regard, our research offers a robust methodology, which helps to advance the development of scientific theories and knowledge reproduction in our field.

Acknowledgments Some of the material in this chapter previously appeared in the paper "Using Decision Tree Modelling to Support Peircian Abduction in IS Research: A Systematic Approach for Generating and Evaluating Hypotheses for Systematic Theory Development," ***Information Systems Journal 21:5***, 407–440 (2011).

Appendix A

A Procedure for Implementing the Methodology

Step 1: Preparation (Researcher)

1. **Identify Dependent variables**: Identify dependent variables (e.g., *Performance*).
2. **Identify Possible Mediator variables**: Identify possible mediator variables (e.g., *Utilization*).
3. **Identify Independent variables**: Identify independent variables (e.g., *System Quality*).
4. **Specify the Discretization Method**: An example of such a method is presented in Table 1.
5. **Identify the Target Events for the Dependent and Mediator Variables**: For each dependent variable, identify target events (e.g., *Performance* is High ≡ the value of *Performance* is in the [5.5–7] interval) that may be of interest. Do similarly for the mediator variable(s).
6. **Discretize All Ordinal Variables**: For each ordinal variable, discretize the given variable using the specified discretization method.
7. **Specify DT Generation Parameters**: Specify relevant values for data partitioning parameters (e.g., distribution of cases in training, validation, and test datasets; stratification variables), DT induction parameters (e.g., splitting method options, minimum observations per leaf).
8. **Specify Thresholds**: Specify α, the significance level for statistical testing of the hypotheses, and τ_0, the threshold for p_0. A Strong Single Rule Hypothesis will not be generated for any rule for which the highest p_0 that is supported by the data is below τ_0.
9. **Specify the Set of Decision Rules for Abducting and Evaluating Sibling Rules Hypotheses**: A candidate sibling rules hypothesis is formulated based on the relative frequency distribution of the target event for the set of sibling rules

that are associated with the predictor variable. Appendix B provides an example of a Set of Decision Rules for Formulating and Evaluating Candidate Sibling Rules Hypotheses.

Step 2: Hypotheses Generation (Automatic)
For each dependent variable:

- **Substep 2a: Generate DTs for the given Dependent Variable**: Generate a set of DTs using the discretized dataset and the combination of DT generation parameter values that were specified in *Step 1*.
- **Substep 2b: Abduct and Evaluate Sibling Rules Hypotheses for the given Dependent Variable:** In this substep, for each DT, a sibling rules hypothesis will be abducted for each set of sibling rules if the relationship between the associated relative frequencies is included in the *Set of Decision Rules for Formulating and Evaluating Candidate Sibling Rules Hypotheses*. Each such abducted Sibling Rules Hypothesis is indirectly evaluated by subjecting the associated set of surrogate hypotheses to statistical testing. If they are all accepted, then there is good reason to believe that the Sibling Rules Hypothesis will not be rejected, and so it is not rejected.
- **Substep 2c: Abduct Strong Single Rule Hypotheses for the given Dependent Variable**: For each sibling rule that is statistically different from each of its other sibling rules at significance level α, determine whether it is a strong rule by testing the surrogate hypothesis: $p_0 \geq \tau_0$. For each such surrogate hypothesis that is accepted, use the given rule to abduct a corresponding Strong Single Rule Hypothesis.

Step 3: Identify Mediator Variables (Automatic)

- Examine the set of supported hypotheses from Step 2 to determine whether any potential mediator variable is included at least one of the abducted hypotheses for one of the dependent variables.

Step 4: Generate Hypotheses for Mediator Variables (Automatic)
Step 4 is executed for each mediator variable that was included in a condition event of a hypothesis for one of the dependent variables. If there is no such mediator variable, then this step is bypassed.

For each mediator variable:

- **Substep 4a: Generate DTs for Predicting the given Mediator Variable**. Similar to Substep 2a.
- **Substep 4b: Abduct Sibling Rules Hypotheses for Mediator the given Variable.** Similar to Substep 2b.
- **Substep 4c: Abduct Strong Single Rule Hypotheses for the given Mediator Variable.** Similar to Substep 2c.

Step 5: Abduction of Theoretical Model (Automatic)
Generate a theoretical model by integrating the set of causal links between predictor and target variables that are associated with the abducted directional and Single Rule Hypotheses.

Appendix B

Examples of Decision Rules for Formulating and Evaluating

Candidate Sibling Rules Hypotheses

Assuming a discretized target variable with 3 bins, Table 9 could be used to formulate and test a candidate sibling rules hypothesis of the form: *Predictor variable X has {***Impact Type***} on Target variable Y.*

Table 9 Example of decision rules for formulating candidate sibling rules hypotheses

Impact type	Relationships between relative frequencies	Set of surrogate hypotheses that must each be accepted
A positive impact	$f_H > f_M > f_L$	$(p_H > p_M)$ & $(p_M > p_L)$
A positive impact	$f_H > f_{L\_M}$	$p_H > p_{L\_M}$
A negative impact	$f_H < f_M < f_L$	$(p_H < p_M)$ & $(p_M < p_L)$
A negative impact	$f_H < f_{L\_M}$	$p_H > p_{L\_M}$
U-shaped symmetric curvilinear impact	$(f_L > f_M)$ & $(f_H > f_M)$	$(p_L > p_M)$ & $(p_H > p_M)$
Inverted U-shaped symmetric curvilinear impact	$(f_L < f_M)$ & $(f_H < f_M)$	$(p_L > p_M)$ & $(p_H > p_M)$
No impact		$(p_L \approx p_M)$ & $(p_M \approx p_H)$
No impact		$p_H \approx p_{L\_M}$

References

Benbasat I, Zmud R (1999) Empirical research in information systems: the practice of relevance. MIS Q 23(1):3–16

Brusic V, Zeleznikow J (1999) Knowledge discovery and data mining in biological databases. Knowl Eng Rev 14:257–277

Chalmers AF (1994) What is this thing called science? 3rd edn. Hackett Publishing

Dipert R (1995) Peirce's underestimated role in the history of logic. In: Ketner K (ed) Peirce and contemporary thought. Fordham University Press, New York

Doll WJ, Torkzadeh G (1988) The measurement of end-user computing satisfaction. MIS Q 12(2):259–274

Etezadi-Amoli J, Farhoomand AF (1996) A structural model of end user computing satisfaction and user performance. Inf Manage 30(2):65–73

Fann KT (1970) Peirce's theory of abduction. Martinus Nijhoff, Amsterdam

Grimes TR (1990) Truth, content, and the Hypothetico-Deductive method. Philosophy of Science, 57, 514–522

Goodhue DL, Thompson RL (1995) Task-technology fit and individual performance. MIS Q 19(2):213–236

Hanson NR (1961) Is there a logic of discovery. In: Feigle H, Maxwell G (eds) Current issues in the philosophy of science. Holt, Rinehart and Winston, pp 20–35

Harman G (1965) Inference to the best explanation. Philos Rev 74:88–95
Hintikka J (1968) The varieties of information and scientific explanation. In: van Rootselaar B, Staal JF (eds) Logic, methodology and philosophy of science III. North Holland, pp 151–171
Hintikka J (1997) The place of CS Peirce in the history of logical theory. In: Lingua Universalis vs Calculus Ratiocinator, selected papers 2. Kluwer, pp 140–161
Kim H, Koehler G (1995) Theory and practice of decision tree induction. Omega 23(6):637–652
Ko M, Osei-Bryson K-M (2004a) Exploring the relationship between information technology investments and firm performance productivity using regression splines analysis. Inf Manage 42:1–13
Ko M, Osei-Bryson K-M (2004b) Using regression splines to assess the impact of information technology investments on productivity in the health care industry. Inf Syst J 14(1):43–63
Lee C, Irizarry K (2001) The GeneMine system for genome/proteome annotation and collaborative data mining. IBM Syst J 40(2):592–603
Niiniluoto I (1993) Peirce's theory of statistical explanation. In: Moore EC (ed) Charles S Peirce and the philosophy of science. The University of Alabama Press, Tuscaloosa, pp 186–207
Niiniluoto I (1999) Defending abduction. Proc Philos Sci 66:S436–S451
Palys TS (2003) Research decisions: quantitative and qualitative perspectives, 3rd edn. Nelson, Scarborough
Popper KR (1957) The aim of science. Ratio 1
Popper K (1963) Conjectures and refutations: the growth of scientific knowledge. Routledge and Kegan Paul, London, UK
Popper KR (1968) The logic of scientific discovery. Harper Torch Books, New York
Putnam H (1982) Peirce the logician. Historia Math 9:290–301
Quine WV (1995) Peirce's Logic. In: Ketner KL (ed) Peirce and contemporary thought. Fordham, New York, pp 23–31
Quinlan JR (1986) Induction of decision trees. Mach Learn 1:81–106
Tursman R (1987) Peirce's theory of scientific discovery. Indiana University Press, Bloomington

Chapter 5
Application of a Hybrid Induction-Based Approach for Exploring Cumulative Abnormal Returns

Francis Kofi Andoh-Baidoo, Kwasi Amoako-Gyampah and Kweku-Muata Osei-Bryson

1 Introduction

In various business disciplines including finance, accounting, information systems, and operations management, there is an interest in determining whether the public announcement of a business-related event has a statistically significant impact on the market value of the firm. For example, several researchers have explored the characteristics of Internet security breaches on the market value of the breached firms (e.g., Campbell et al. 2003; Cavusoglu et al. 2004; Gordon and Loeb 2002; Hovav and D'Arcy 2004). The relevant Capital Market reaction is typically assessed based on cumulative abnormal return (CAR), which is the sum of abnormal returns over the event window. In other words, 'The abnormal return is the actual ex post return of the security over the event window minus the normal return of the firm over the event window' (Mackinlay 1997, p. 15). A popular approach for exploring the occurrence of CAR involves the application of the event study methodology. The event study methodology typically has two goals: (1) to determine whether or not an event, such as the announcement of Internet security breach, or an e-commerce initiative leads to CAR and (2) to examine the factors that influence the observed CAR.

F. K. Andoh-Baidoo (✉)
Department of Computer Information Systems and Quantitative Methods,
The University of Texas—Pan American, Edinburg TX 78539, USA
e-mail: andohbaidoof@utpa.edu

K. Amoako-Gyampah
Bryan School of Business and Economics, The University of North Carolina at Greensboro,
Greensboro NC 27402-6170, USA
e-mail: k_amoako@uncg.edu

K.-M. Osei-Bryson
Department of Information Systems, Virginia Commonwealth University,
301 W. Main Street, Richmond VA 23284, USA
e-mail: KMOsei@VCU.Edu

K.-M. Osei-Bryson and O. Ngwenyama (eds.), *Advances in Research Methods for Information Systems Research*, Integrated Series in Information Systems 34,
DOI: 10.1007/978-1-4614-9463-8_5,

While both confirmatory (e.g., regression analysis) approaches and exploratory approaches (e.g., decision tree (DT) induction) can be used to analyze statistical data such as those involving predictors of CAR, most previous event studies have used confirmatory approaches such as regression and ANOVA, in examining the determinants of CAR. Confirmatory approaches offer several advantages. However, they require the prior explicit specification of all hypotheses. This involves the reliance on strengths and limitations of human imagination for the generation of the hypotheses, an imagination that is typically developed based on the insights and limitations from previous studies. However, in many situations, it could be difficult for the researcher to identify every relevant hypothesis that might be testable, particularly the local (conditioned) hypotheses. The researcher may not be able to discover additional important relationships in the data among the variables that are not explicitly specified in the postulated hypotheses. In this chapter, we discuss the application of an exploratory approach that involves the use of DT induction, a data mining technique. This method is based on the approach to hypotheses generation that was presented in Osei-Bryson and Ngwenyama (2011).

The DT induction technique is described in detail elsewhere in this book. DT induction is used to partition the dataset into subsets based on input variables selected by the relevant splitting method. DT induction identifies those variables that are significant in predicting the outcome with the most significant attribute located at the root of the tree, while succeeding attributes further discriminate between the outcomes. In this chapter, we use two previous studies to demonstrate the applicability of the hybrid methodology of DT induction and nonparametric analysis in examining the impact of two information system events (Andoh-Baidoo et al. 2010, 2012).

Our major motivations for suggesting this approach are that DTs provide an interpretable model in the form of understandable and actionable rules that may be used by decision-makers and that a DT-based solution may provide additional insights beyond what may be provided by regression. The approach presented here is very useful where prior studies present mixed results and where there are concerns about return variability in the case of market return studies (Andoh-Baidoo et al. 2012). Thus, the approach presented in this chapter can enable a more comprehensive understanding of business phenomena without specific a priori statements about the directions and magnitudes of the variables that might influence the phenomena but allow the data analysis process to guide the formulation of relationships that exist between the variables.

The rest of the chapter has the following sections: an overview on the event study methodology; a description of our DT induction-based method; and illustrations of the DT-based method.

2 The Event Study Methodology

The event study methodology is based on the efficient market hypothesis, which posits that capital markets are efficient mechanisms to process information available about firms (Fama et al. 1969). Specifically, the event study methodology is the test for the semi-strong form of the efficient market hypothesis. This form of hypothesis states that investors process publicly available information about the activities of a firm that impact the firm's current and future performance. Further, as new information about the firm's activities that can potentially affect the firm's future earnings is publicized, the stock price changes relatively quickly to reflect the current assessment of the value of the firm. A positive announcement is expected to lead to positive CAR. The CAR represents the excess return observed over a period during which an announcement is made over normal return that is expected in the absence of the event. The relevance of the event studies is noted in this way: 'The event study methodology has…become the standard method of measuring security price reaction to some announcement' (Binder 1998, p. 111).

Since the seminal works of Ball and Brown (1968) and Fama et al. (1969), the event study methodology has found application in a wide range of issues in the information system discipline (e.g., Aggarwal et al. 2006; Dos Santos et al. 1993; Dow et al. 2006; Guan et al. 2006; Kannan et al. 2007; Subramani and Walden 2001), as well as other disciplines including operations management and finance (Bodie et al. 2001; Brown and Warner 1980; Corrado 1989; Cowan 1992; Hendricks et al. 1995).

The main goal of the event study methodology is to determine whether an event such as the announcement of a newly created CIO position leads to CAR. The event study methodology has been used to show that, beyond the firm's past financial performance data, other factors influence the observed CAR resulting from the announcements of some events in the public media. Firms would be interested in knowing which combinations of firm and event characteristics determine whether the event would lead to positive or negative CAR. Hence, the second goal in event studies has been to examine what predictor variables influence abnormal returns and the magnitude of the effect.

The use of DT induction has been found to provide additional insights on the conditional relationships between independent and dependent variables that may not have been established in postulated hypotheses (Osei-Bryson and Ngwenyama 2011). Further, the use of data mining for data analysis presents additional insights that might not be detected by the regression models alone (Ko 2003; Murphy 1998; Osei-Bryson and Ko 2004). For example, in the study by Murphy (1998), 'Analysis with linear regression identified only one significant attribute…the induced DTs revealed useful patterns…' (p. 189).

Most traditional event studies have used parametric tests for the first goal, while regression analysis is used for the second goal (Cavusoglu et al. 2004; Subramani and Walden 2001). Details of the traditional event study methodology can be found in studies such as Binder (1998) and Mackinlay (1997). The DT induction-based hybrid methodology involves the use of (1) nonparametric analysis to examine whether an announcement leads to CAR and (2) DT induction to examine the likelihood that some event and firm characteristics may influence the observation of CAR.

The following are the specific activities that differentiate the approach presented in this paper from the traditional approach: (1) use of nonparametric analysis to compute market returns, (2) use of DT induction to examine the relationship between predictor variables and CAR, and (3) presentation of a set of hypotheses before data collection in traditional approach compared with the development of hypotheses after data analysis in the proposed approach.

3 Description of the Hybrid Induction-Based Approach

Figure 1 is a representation of the DT induction-based approach for event studies. The first five steps (in italics) in the proposed approach are similar to those of the traditional event study methodology. The first step in the event study methodology is to identify an event of interest. The second step involves determining the event date. Some events, such as the announcement of the creation of a CIO position, have a single event date. However, in the case where the researcher seeks to understand the impact of the announcements of an event such as an e-commerce initiative, there will be several event dates since such announcements will involve several firms over a period of time. The predictor variables are also defined most likely[1] in the third stage. In the fourth stage, the researcher defines the event window, which is the period that the researcher seeks to examine the impact of the event. Generally, the event window is defined to be larger than the specific period of interest but should be short relative to the estimation period. A typical event window is 3 days covering a day before the announcement through the day after the announcement. The estimated window is the period over which the normal stock market return is estimated. Typically, this period is 120 days, but 160 days has been used in some studies. Generally, the event period is not included in the estimation window to prevent the event from influencing the normal performance model. The data collection is the next stage. Using specific criteria for identifying the event, all data on the event are collected over the desired period in a selected public media and coded based on a set of criteria defined by the researcher according to the specific characteristics of the event and predictor variables.

3.1 Estimating Parameters of Event-Generating Model and Computing CAR

In order to compute CAR and average CAR, first, the normal return in the absence of the event is computed. Two common approaches employed in the estimation of

[1] In some cases, steps 3 and 4 may be interchanged. In the traditional event study approach, step 3 may precede all the previous steps and may even be the reason for the research project in general. This is because the research interest in understanding how the predictor variables may influence CAR may be the reason for the entire project in the first place.

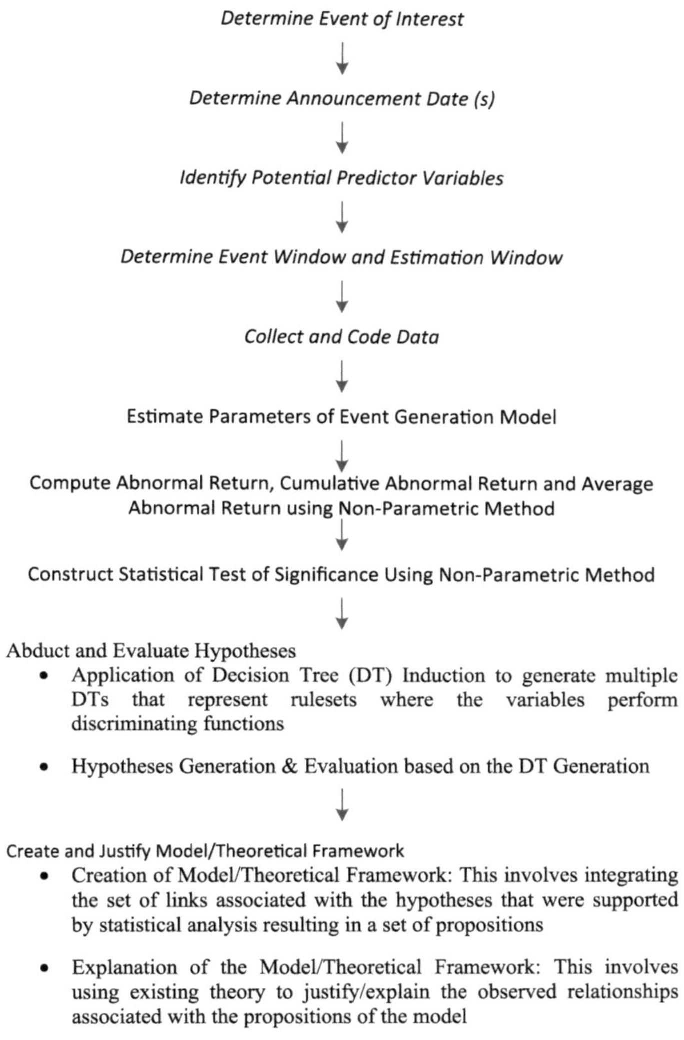

Fig. 1 Hybrid DT induction-based event studies approach

normal returns are as follows: the Constant Mean Return Model and the Market Model; the Market Model is the most frequently used.

Based on the Market Model (Sharpe 1963), the return of a specific stock can be represented as

$$R_{it} = \alpha_i + \beta_i R_{mt} + \varepsilon_{it}$$

where R_{it} = return of stock i on day t; R_{mt} is the return of the market portfolio on day t, α_i and β_i are the intercept and slope parameters, respectively, for firm i, and ε_{it} is the disturbance term for stock i on day t.

The abnormal return for firm i on day t of the event window is computed as

$$AR_{it} = R_{it} - \left(\hat{\alpha}_i + \hat{\beta}_i R_{mt}\right)$$

where $\hat{\alpha}$ and $\hat{\beta}$ are the ordinary least square estimates of α and β. These parameters are estimated using the Market Model over the estimation window: the 120-day period ending with the day immediately preceding the first day of the event window, i.e., day (−2).

The summation of the daily abnormal returns over the event window is the CAR. The CAR for stock i over the event window ($T1$, $T2$) is computed as

$$\text{CAR}_{i(T1,T2)} = \sum_{t=T1}^{T2} \text{AR}_{it}$$

For a sample of n stocks, the average CAR over the event window is

$$\text{CARR}_{(T1,T2)} = \frac{1}{n} \sum_{i=1}^{n} \text{CAR}_{i(T1,T2)}$$

3.2 Test of Significance: Nonparametric Analysis

Nonparametric tests have been shown to provide higher power in the detection of abnormal returns than traditional parametric tests (Corrado 1989; Cowan 1992). Brown and Warner (1985) show that parametric tests report 'false' price more often than nonparametric tests when there are event-related variances. The advantage of nonparametric tests over parametric tests is that the nonparametric statistic is not subjected to stringent assumptions about return distributions as a parametric test does.

Among the nonparametric tests that have been used in event studies are the generalized sign test (Cowan 1992), the rank test (Cowan 1992), and the time series standard deviation method (Brickley et al. 1991; Dopuch et al. 1986). 'The generalized sign test examines whether the number of stocks with positive CARs in the event window exceeds the number expected in the absence of abnormal performance' (Cowan 1992, p. 5). The rank test treats the 120-day estimation period and the event day as a single 120-day time series and assigns a rank to each daily return for each firm. Although the rank test is more powerful than the generalized sign test, in the case where the return variance increases, the generalized sign test offers the better choice (Cowan 1992). In particular, Cowan shows that when a single stock in a portfolio has extreme positive return, the generalized sign test is correctly specified, while the rank test is not. Similarly, although the time series standard deviation method computes a single variance estimate for the entire portfolio without consideration of the unequal return variances across firms or events, it evades the potential problems of cross-sectional correlation of security returns. We used the generalized sign test to check for significance of CAR in our illustrative examples.

3.3 Hypotheses Abduction and Evaluation

Given a set of events that have been classified by some kind of dichotomous or nominal categorical variables with one of the variables assigning an event as either abnormal (positive[2] CAR) or normal (zero or negative CAR) based on other predictor variables, we can use DT induction to generate a set of rules that can be employed to assign new events as abnormal or normal. These set of rules and outcomes can provide understanding of the relationships between CAR and the predictor variables. While regression analysis determines how much the specific variables influence the level of CAR, DT induction is used to measure the likelihood that the event would lead to positive CAR or negative CAR. Using the rules developed by DT induction sets of hypotheses can be generated and then tested using traditional statistical analysis.

3.4 Creation of Model/Theoretical Framework

Finally, a set of propositions can be created from the hypotheses developed during the hypotheses abduction and evaluation step. These hypotheses form a theoretical model that can further be analyzed using a traditional deductive theory development approach. Hence, our proposed approach provides new ways to develop both inductive and deductive theories.

4 Illustrative Examples

Here, we illustrate two applications of the hybrid DT-based approach in the literature. The following discussion is for illustrative purposes; a complete description of the application of the methodology described in this chapter on electronic commerce initiatives and Internet security breaches is presented elsewhere (Andoh-Baidoo et al. 2010, 2012).

4.1 Example 1: e-Commerce Announcements and CAR

In the first step, the researchers were interested in examining the effect of the announcements of electronic commerce initiatives on the market value of the firms

---

[2] For an event such as Announcement of Security Breaches, *abnormal* represents CAR with negative values and *normal* represents CAR with zero or positive value.

Table 1 Cumulative abnormal return for the 3-day window

Days	Cumulative average abnormal return Equally weighted (%)	Z	Positive:normal	Generalized sign Z
(−1, +1)	1.83	5.844***	497:449	4.085***

***significant at 0.001

making the announcements. Hence, the event of interest is the announcement of an electronic commerce initiative. In the second step, the researchers selected the period 1997–2003 to capture the event for both the pre- and post-Internet bubble eras. In step 3, predictor variables selected for the study were based on prior work on CAR and e-commerce initiatives. In step 4, a 3-day event window was selected based on the literature. This covers a day before the announcement, the day of the announcement, and a day after the announcement. A 120-day estimated window was used.

In step 5, data were collected from the PR Newswire and the Business Wire. The criteria used to eliminate 'unwanted events' were as follows: (1) only the first announcement were selected if a single event were reported multiple times in a single source or multiple sources; (2) only firms that were listed on the exchange from which market parameter estimates could be obtained and those firms listed in the specific database where stock prices could be obtained were used; (3) only firms included in the research database and had returns available for at least a period equal to the estimation window; and (4) where there were confounding effects such as earning announcements, dividends, or any major announcement in the event window involving a firm included in the sample, the event was dropped. In step 5, a content guide that provided a description of the two different categories for each predictor variable was used to code each event.

Following the data collection and coding, Eventus® software was used in the computation of abnormal return. Eventus® effectively interfaces with SAS and CRSP to generate test results. Table 1 is a sample result from the study (Andoh-Baidoo et al. 2012). Within the 3-day event window, the ratio of positive to normal returns was greater than 1 (497:449) with a corresponding Z value that indicates that the CAR for the 3-day event window (a day before the announcement through a day after the announcement) was positive and significant at the 1 % level. This suggests that announcements of e-commerce initiatives in the public media lead to positive CAR. For the 3-day event window, the average CAR for the 946 events was about 1.83 %.

4.1.1 Hypothesis Abduction and Evaluation

DT induction was used to partition the dataset into subsets based on input variables selected by the relevant splitting method. In a DT, nodes that have the same non-root parent node (i.e., input variable) are referred to as sibling nodes, where

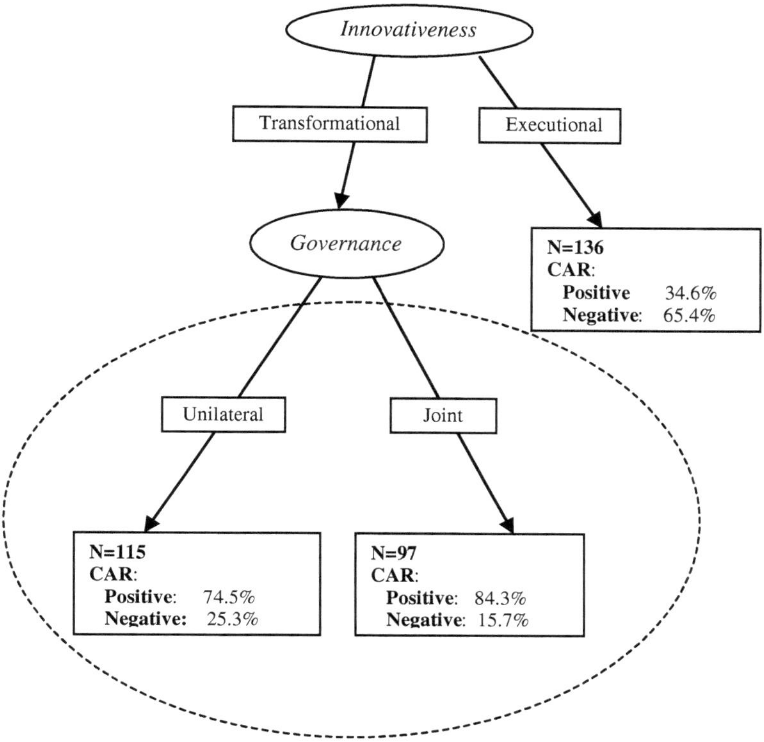

Fig. 2 Set of sibling rules with *Governance* as the subject variable

each sibling is associated with a mutually exclusive subset of the values of the relevant input variable and the relevant value of any higher ancestor node.

Our hypotheses were created in one of two ways: (1) from a single pair of sibling rules, which results in the first-order sibling rule hypothesis; or (2) from two pairs of sibling rules, which results in the second-order sibling rule hypothesis.

4.1.2 First-Order Sibling Rule Hypothesis

A first-order sibling rule hypothesis is based on one set of sibling rules. Consider the pair of sibling rules (see Fig. 2) where all conditions are the same (Innovativeness is Transformational) except for the one involving the given discriminating variable (e.g., Governance):

- IF *Innovativeness* is *Transformational* and *Governance* is *Unilateral* THEN *CAR* is *Positive* with probability 74.5 % and N (i.e., number of Cases) $= 115$;
- IF *Innovativeness* is *Transformational* and *Governance* is *Joint* THEN *CAR* is *Positive* with probability 84.3 % and $N = 97$.

The existence of this pair of sibling rules leads to the creation of the hypothesis: 'IF *Innovativeness* is *Transformational* THEN *Governance* is a predictor of CAR.' Governance becomes a discriminating predictor in this case. For the given target event (e.g., CAR is positive), the posterior probabilities for each sibling node are compared. If for any pair of sibling nodes, the relevant posterior probabilities are very different, then this would suggest that the given variable is a predictor for the target event (Osei-Bryson and Ngwenyama 2004). In this manner, a given set of sibling rules can be used to generate and test hypotheses that involve conjecturing that the given variable is a predictor of CAR. If the number of cases associated with a given set of sibling nodes is sufficiently large, then the hypothesis may be subjected to statistical analysis. Table 2 presents examples of first-order sibling rule. A proportion test is performed to investigate whether the difference in posterior probabilities (proportions or relative frequencies of the number of cases that are positive), i.e., a test at the 5 % level, for the sibling nodes of the discriminating variable is the same and that the difference did not occur by chance. The difference is between two population proportions ($\hat{P}_1 - \hat{P}_2$) based on two independent samples of size n_1 and n_2 with sample proportions $\hat{P}_1$ and $\hat{P}_2$.

The test statistic is given by

$$Z = \frac{\hat{P}_1 - \hat{P}_2}{\sqrt{\frac{\hat{P}_1(1-\hat{P}_1)}{n_1} + \frac{\hat{P}_2(1-\hat{P}_2)}{n_2}}}$$

4.1.3 Second-Order Sibling Rule Hypothesis

A second-order sibling rule hypothesis (see Table 3) is based on two sets of sibling rules (say S_1, S_2) that have the following conditions:

Statistical testing of a second-order sibling rules hypothesis involves

$$Z = \left(\left(\rho_{11,21} - \rho_{11,22}\right) - \left(\rho_{12,21} - \rho_{12,22}\right)\right)/s_p^{1/2}$$

where

$$\begin{aligned} s_p = &\left(\rho_{11,21}(1 - \rho_{11,21})/n_{11,21} + \rho_{11,22}(1 - \rho_{11,22})/n_{11,22} \right.\\ &\left. + \rho_{12,21}(1 - \rho_{12,21})/n_{12,21} + \rho_{12,22}(1 - \rho_{12,22})/n_{12,22}\right) \end{aligned}$$

Table 4 is an example of second-order sibling rule hypothesis.

4.1.4 Creation of Model/Theoretical Framework

The framework involves integrating the set of causal links associated with the abducted hypotheses that were supported by statistical analysis. It describes the independent and dependent variables and the newly hypothesized relationships that were supported. This resulting framework can be presented in terms of

Table 2 Examples of first-order sibling rule hypothesis

	Moderating condition	Discriminating condition	CAR relative frequency	*N*
Pair of sibling rules	None	*Innovativeness* = 'TRANSFORMATIONAL'	0.79	212
		Innovativeness = 'EXECUTIONAL'	0.346	136
Hypothesis	Transformational *Innovativeness* is more favorable than executional *innovativeness*			
	Accepted (p value <0.001)			
Pair of sibling rules	*Innovativeness* = 'TRANSFORMATIONAL'	Governance = 'UNILATERAL'	0.745	115
		Governance = 'JOINT'	0.843	97
Hypothesis	If *Innovativeness* is 'TRANSFORMATIONAL,' JOINT *Governance* is more favorable than UNILATERAL *Governance*			
	Accepted (p value = 0.05)Pair of sibling rules			

Table 3 Format of second-order sibling rule hypothesis

Set	Rule	Moderating condition	Discriminating conditions		Relative frequency of CAR	N	Diff
			Variable 1	Variable 2			
S_1	R_{11}	Same for all rules	DC_{11}	DC_{21}	$\rho_{11,21}$	$n_{11,21}$	$\rho_{11,21} - \rho_{11,22}$
	R_{12}		DC_{11}	DC_{22}	$\rho_{11,22}$	$n_{11,22}$	
S_2	R_{21}		DC_{12}	DC_{21}	$\rho_{12,21}$	$n_{12,21}$	$\rho_{12,21} - \rho_{12,22}$
	R_{22}		DC_{12}	DC_{22}	$\rho_{12,22}$	$n_{12,22}$	
Hypothesis			Given {moderating condition}s, the difference in CAR between DC_{21} and DC_{22} is greater for DC_{11} than for DC12.				

Note
DC_{11}: Condition 1 of discriminating variable 1; DC_{12}: Condition 2 of discriminating variable 1
DC_{21}: Condition 1 of discriminating variable 2; DC_{22}: Condition 2 of discriminating variable 2

Table 4 Example of second-order sibling rule hypothesis without moderator

Set	Rule	Moderating condition	Discriminating conditions		Relative frequency of CAR	N	Diff
			Customer type	Innovativeness			
S_1	R_{11}	None	B2C	Transformational	0.69	134	+0.26
	R_{12}		B2C	Executional	0.43	315	
S_2	R_{21}	None	B2B	Transformational	0.82	245	+0.55
	R_{22}		B2B	Executional	0.27	252	
Hypothesis	The increase in CAR between transformational and executional *innovativeness* is greater for B2B than B2C						

Table 5 Sample propositions from first- and second-order sibling rule hypotheses

Variable	Proposition
First order	
Innovativeness	Firms are more likely to experience positive abnormal returns for e-commerce announcements involving transformational than executional innovativeness
Governance	Firms that make e-commerce announcements are more likely to experience positive abnormal returns for joint governance structure than unilateral governance structure if the initiative is transformational
Second order	
Product type innovativeness	The increase in CAR between Transformational and executional *innovativeness* is greater for *digital* products than for *tangible* products
Customer type time	For transformational *innovativeness*, the increase in CAR before and after March 2000 Internet bubble crash is greater for B2C than for B2B

interactions between the predictor variables in the determination of CAR and corresponding propositions (see Table 5 for sample propositions for both first-order and second-order sets of hypotheses). The results of the inductive and abductive approaches presented above allow us to make propositions (see Table 5) about CAR and the predictor variables that can be deductively tested using different datasets to provide additional insights. Hence, existing theories can be enhanced, or new theories can be developed.

4.2 Example 2: Internet Security Breaches and CAR

In illustrative example 2, the event of interest was the announcement of Internet security breaches in the public media. A complete description of this example can be found in (Andoh-Baidoo et al. 2010). 1997–2003 was selected as the period of interest. Predictor variables were based on Howard's study (Howard 1997). Just as in the previous example, a 3-day event window was used for the Internet security

Table 6 Cumulative abnormal returns for Internet security breach sample

Days	Cumulative average abnormal return		Median cumulative abnormal return (%)	Z	Positive: negative	Generalized sign Z
	Equally weighted (%)	Precision weighted (%)				
(−1, +1)	−3.18	−1.75	−1.45	−1.94*	14:27	−1.72*

*significant at 0.05

Table 7 Sample set of sibling rule hypothesis

	Moderating condition	Discriminating condition	CAR relative frequency	*N*
Pair of sibling rules	Access = 'UNAUTHORIZED ACCESS'	Period = 'POST FEB 2000'	0.83	24
		Period = 'PRE FEB 2000'	0.50	8
	If Access is 'UNAUTHORIZED USE,' *Net* Firm is more likely to lead to negative CAR than *Non-Net* Firm			
Hypotheses	***Accepted p value = 0.04			
Pair of sibling rules	Period = 'POST FEB 2000' and Access = 'UNAUTHORIZED ACCESS'	Breach Objective = 'CHALLENGE/ STATUS' or 'FINANCIAL GAIN'	1.00	12
		Breach Objective = 'DAMAGE' or 'POLITICAL GAIN'	0.67	12
Hypothesis	If Period is Post Feb 2000 and Access is Unauthorized Access, Attack with Objective to challenge vulnerability status or financial gain is more likely to lead to negative CAR than Objective to cause Damage or seek Political Gain			
	***Accepted p value = 0.007			
Variable	Proposition			
Period	Firms are more likely to experience negative abnormal returns for Internet security breach announcements for Post Feb 2000 than Pre Feb 2000 periods			
Objective	For Post 2000 periods, firms are more likely to experience negative abnormal returns for Internet security breaches if the objective of the attack is either for financial gain or challenge/status than when the objective is to cause damage or for political gain			

breach event. Data were collected from the Wall Street Journal, New York Times, Financial Times, Washington Post, and USA Today. Using the same criteria as described for the e-commerce event, unsuitable events were eliminated.

Following the previous example, using SAS, CRISP data, and Eventus® software, Table 6 shows the CAR for the 3-day event window. A firm loses approximately 3.18 % of its market value when it makes a public announcement that it has been breached. Here, median CAR and precision-weighted CAR were also recorded. Although all three are different, each indicates decrease in CAR due to the announcement of an Internet security breach in the public media.

Table 7 presents examples of first-order sibling rule hypotheses for the Internet security breach event. Details of the predictor variables are presented in (Andoh-Baidoo et al. 2010).

5 Conclusion

In this chapter, we have discussed how DT induction can be used in conjunction with a nonparametric test in event studies. We have illustrated how the approach has been employed to examine two information system events, e-commerce initiative announcements, and Internet security breach announcements affect CARs. The approach can be employed in other domains. The approach enables the development of hypotheses that can be tested statistically.

Acknowledgments Some of the materials in this chapter previously appeared in 'Effects of Firm and IT Characteristics on the Value of e-Commerce Initiatives: An Inductive Theoretical Framework,' *Information Systems Frontiers 14*:2, 237–259 (2012).

References

Aggarwal N, Dai Q, Walden EA (2006) Do markets prefer open or proprietary standards for XML standardization? Int J Electron Commer 11(1):117–136

Andoh-Baidoo FK, Amoako-Gyampah K, Osei-Bryson K-M (2010) How internet security breaches harm market value. IEEE Secur Priv 8:36–42

Andoh-Baidoo FK, Osei-Bryson K-M, Amoako-Gyampah K (2012) Effects of firm and IT characteristics on the value of e-commerce initiatives: an inductive theoretical framework. Inf Syst Front 14:237–259

Ball R, Brown P (1968) An empirical evaluation of accounting income numbers. J Acc Res 6:159–178

Binder JJ (1998) The event study methodology since 1969. Rev Quant Finance Acc 11(2):111–137

Bodie Z, Kane A, Marcus AK (2001) Essentials of investments, 4th edn. McGraw-Hill, Boston

Brickley JA, Dark FH, Weisbach MS (1991) The economic effects of franchise termination laws. J Law Econ 34(1):101–132

Brown S, Warner J (1980) Measuring security price performance. J Financ Econ 8:205–250

Brown S, Warner J (1985) Using daily stock returns: the case of event studies. J Financ Econ 14:3–31

Campbell K, Gordon LA, Loeb MP, Zhou L (2003) The economic cost of publicly announced information security breaches: empirical evidence from the stock market. J Comput Secur 11(3):431–448
Cavusoglu H, Mishra B, Raghunathan S (2004) The effect of Internet security breach announcements on market value: capital market reactions for breached firms and Internet security developers. Int J Electron Commer 9(1):69–104
Corrado CJ (1989) A nonparametric test for abnormal security-price performance in event studies. J Financ Quant Anal 25:549–554
Cowan AR (1992) Nonparametric event study tests. Rev Quant Financ Acc 2:343–358
Dopuch N, Holthausen RW, Leftwich RW (1986) Abnormal stock returns associated with media disclosures of 'subject to' qualified audit options. J Acc Econ 8(2):93–117
Dos Santos BL, Peffers K, Mauer DC (1993) The impact of information technology investment announcement on the market value of the firm. Inf Syst Res 4(1):1–23
Dow KE, Hackbarth G, Wong J (2006) Enhancing customer value through IT investments: a NEBIC perspective. Database Adv Inf Syst 37(2&3):167–175
Fama EF, Fisher L, Jensen MC, Roll R (1969) The adjustment of stock prices to new information. Int Econ Rev 10(1):1–21
Gordon LA, Loeb MP (2002) The economics of information security investment. ACM Trans Inf Syst Secur 5(4):438–457
Guan L, Sutton SG, Chang CJ, Arnold V (2006) Further evidence of shareholder effects of announcements of newly created CIO positions. Database Adv Inf Syst 37(2&3):176–187
Hendricks KB, Singhal VR, Weidman CI (1995) The impact of capacity expansion on the market value of the firm. J Oper Manage 12:259–272
Hovav A, D'Arcy J (2004) The impact of virus attack on the market value of firms. Inf Syst Secur J 13(3):32–40
Howard J (1997) An analysis of security incidents on the internet. Unpublished PhD Thesis, Carnegie Mellon University
Kannan K, Rees J, Sridhar S (2007) Market reactions to information security breach announcements: an empirical analysis. Int J Electron Commer 12(1):69–91
Ko MS (2003) An exploration of the impact of information technology investment on organizational performance. Virginia Commonwealth University, Richmond, VA
Mackinlay C (1997) Event studies in economics and finance. J Econ Lit 35:13–39
Murphy CK (1998) Induced decision trees for temporal medical data. Paper presented at the 4th international conference on information systems, Baltimore, MD
Osei-Bryson K-M, Ko MS (2004) Exploring the relationship between information technology investments and firm performance using regression splines analysis. Inf Manage 42:1–13
Osei-Bryson K-M, Ngwenyama OK (2004) Peirce, popper and data mining: an approach to empirically based theory development and testing. Information Systems Research Institute, Virginia Commonwealth University, Richmond, VA
Osei-Bryson K-M, Ngwenyama O (2011) Using decision tree modelling to support peircian abduction in IS research: a systematic approach for generating and evaluating hypotheses for systematic theory development. J Inf Syst 21(5):407–440
Sharpe W (1963) A simplified model for portfolio analysis. Manage Sci 9:277–293
Subramani M, Walden E (2001) The impact of E-Commerce announcements on the market value of firms. Inf Syst Res 12(2):135–154

Chapter 6
Ethnographic Decision Tree Modeling: An Exploration of Telecentre Usage in the Human Development Context

Arlene Bailey and Ojelanki Ngwenyama

This chapter presents an investigation of the decision-making process of community members who decide on using telecentres for entrepreneurial endeavours. A qualitative approach through interviews with telecentre staff and users is used to assess the barriers and enablers to economic activity through use of telecentres. An ethnographic decision tree model (EDTM) is developed to illustrate the process through which a community member makes a decision to use the telecentre to support economic livelihood. A predictive model for entrepreneurial behaviour is proposed based on the factors which influence the usage of telecentres for entrepreneurship—social ties, opportunity recognition and support from the telecentres.

1 Introduction

With the increasing focus on the assessment of the role of information and communication technologies (ICTs) in enhancing social and economic development, it is important to be able to utilize relevant and rigorous research methods which can support these key research endeavours. Many ICTs for development initiatives have been implemented, and there have been calls for more research in this area in an effort to assess the impact of these initiatives (Ngwenyama et al. 2006; Walsham et al. 2007). Given that these initiatives are focused on human development and rely on the usage of technologies, particularly by under-served groups within communities, it is useful for researchers and practitioners to understand people's thoughts and actions in deciding whether to interact with these ICTs and

A. Bailey (✉)
Department of Sociology, Psychology and Social Work, University of the West Indies, Kingston 7, Mona, Jamaica
e-mail: arlene.bailey@uwimona.edu.jm

O. Ngwenyama
Ted Rogers School of Management, Ryerson University, 350 Victoria Street, Toronto, ON M5B 2K3, Canada
e-mail: Ojelanki@ryerson.ca

K.-M. Osei-Bryson and O. Ngwenyama (eds.), *Advances in Research Methods for Information Systems Research*, Integrated Series in Information Systems 34,
DOI: 10.1007/978-1-4614-9463-8_6,

integrate them in their daily lives. The decision tree is a useful way of organizing knowledge used in the decision-making process and classifying resulting decisions, and there are multiple factors that are important in generating and evaluating decision trees (Osei-Bryson 2004).

Ethnographic decision tree modelling (EDTM) (Gladwin 1989) is a method which allows the identification of a target group's decision-making processes through the perspectives of the members of the target group themselves, as experts in their own decision-making process, enabling tailored policy interventions (Gladwin 1989; Gladwin et al. 2002). The application of this method in empirical investigation of ICT for development initiatives is an example of a relatively new advance in research methods for information systems research (Bailey and Ngwenyama 2013).

This chapter illustrates the use of EDTM in information systems research through an exploration of telecentre usage. In the next section, we provide some background on telecentres in the development context. We then discuss the steps involved in the EDTM process and illustrate the process through which community members make decisions to use telecentres. We conclude with implications of the use of this advanced research method for research, policy and practice in the field of information systems.

2 Telecentres and Human Development

At the community level, telecentres have been established as part of development initiatives to facilitate access to and use of ICTs to improve livelihoods. Although denoted by many names such as community multimedia centres, community technology centres or cybercentres, there is general consensus on defining a telecentre as "a physical space that provides public access to ICTs for educational, personal, social and economic development" (Gomez et al. 1999 p. 17). While each telecentre is grounded in a particular social context, they share a common focus on issues such as community development, social inclusion and assisting in identifying employment opportunities for community members (Bailur 2007a; Gomez et al. 1999; Harris et al. 2003; Madon et al. 2009).

It is important for any initiative to be able to assess the impact of the programme based on demand from users, community and national interests so as to better inform policy design and implementation. Within the local context, Jamaica, there is a great need for research on the use of ICTs for development (ICT4D Jamaica 2008).

The related literature involves discussions on telecentre usage and the evaluation of this usage, decision-making processes among telecentre stakeholders and the relationship between telecentres and employment opportunities and economic activity.

It has been noted that further evaluation of telecentres is needed (Sey and Fellows 2009) and that telecentre services may vary based on the level of national development. Hudson (2001) argues that it is important to examine usage patterns and measure the impact of telecentre usage. She outlines several questions that should be considered in evaluating telecentres. These include focusing on the target groups and whether they use the telecentre more or less than anticipated, the suggestions and additional needs that have been identified, and whether the telecentre is used by other groups of users who were not part of the planned target groups. It is also highlighted that it is useful to observe usage patterns, including the number and frequency of users. However, detailed information about usage is usually limited (Ellen 2003).

3 Research Context and Method

In the area of entrepreneurship and telecentres, research has focused on the entrepreneurial skills of telecentre coordinators (Madon 2005; Kuriyan and Ray 2009). Given the limited literature which focuses on the role of telecentres in user entrepreneurship in the development context, an inductive approach was used to build a theoretical understanding from empirical observations (Blaikie 2000; Korsching and Allen 2004). This enabled the use of interviews and observation to derive an empirically grounded explanation of entrepreneurial behaviour in the context of telecentres, through analysis using the EDTM technique. This strategy of inquiry allows the use of various techniques for the collection and analysis of the data (Denzin and Lincoln 2000) and can provide useful insight into information systems research (Kvasny and Keil 2006; Cavaye 1996; Nandhakumar and Jones 1997).

Telecentres in Jamaica have been established through a number of initiatives and partnerships among international agencies such as the United Nations Development Programme (UNDP), the United Nations Educational, Scientific and Cultural Organization (UNESCO), Inter-American Development Bank (IDB), in collaboration with the Government of Jamaica (GOJ), non-profit non-governmental organizations (NGOs), community-based organizations (CBOs) and private sector organizations.

A total of seven telecentres were selected for this study based on a typology of key characteristics of telecentres articulated by Colle (2000). Sites were chosen to represent different operational characteristics, funding and host institutions, thematic focus and geographic area. Table 1 outlines the telecentres which are part of this study and provides information on the data collection methods employed at each telecentre.

Table 1 Outline of data collected

Name of telecentre	Parish	Area	Interviews with staff	Interviews with users	Survey of users
Association of Clubs	Westmoreland	Rural	3	–	–
Bluefields People's Cooperative Association	Westmoreland	Rural	4	4	24
Caribbean Coastal Area Management Foundation	Clarendon	Semi-urban	2	4	14
Container Project	Clarendon	Semi-rural	1	–	–
Eastern Peace Centre	Kingston	Urban	2	3	23
Liguanea Cybercentre	St. Andrew	Urban	2	2	27
The source—August Town	St. Andrew	Urban	3	2	–

EDTM, an inductive method was used to analyse the interview data to determine patterns of decision logic of individuals who chose to use telecentres for entrepreneurial endeavours. The EDTM method enables researchers to explore decisions that a group will make in real-life situations by using ethnographic techniques to explore the decision process from the decision maker's perspective (Gladwin 1989). These models are built from information elicited from interviews with a sample of users and tested on another sample of persons (Ryan and Bernard 2006; Gladwin 1989). The process of building an EDTM entails exploratory data collection, preliminary model creation and model testing (Ryan and Bernard 2000). The following steps outlined by Gladwin (1989, pp. 22–45) provide details and the basis for developing an EDTM: (1) deciding on the decision that is being modelled; (2) deciding on the set of decision alternatives; (3) conducting ethnographic interviews; (4) conducting participant observation; (5) selecting the sample to use in the model; and (6) deciding on the criteria to use in the model. Steps (7) and (8) involve building the composite decision model for the group from the individual decision trees, while "preserving the ethnographic validity of each individual decision model" (p. 39).

A survey was conducted at four of the telecentres in the study and the information was used to refine and carry out preliminary tests on the EDTM. In order to test the model, the decision criteria were used to calculate the success rate of predicting decisions from a sample of telecentre users.

1. **Deciding on the decision that is being modelled**

In Jamaica, the level of unemployment, which is particularly high among the youth, is recognized as one of the main problems facing the country with "wide-ranging economic and social implications" (Anderson and Williams 2008, p. 23). With low levels of school attendance and educational attainment within some sectors of the population, these youth have difficulty in finding employment. Discussions regarding the role of entrepreneurship in the local economy, particularly for unattached youth, and the policies that should be in place to support these ventures among the self-employed and micro-, small and medium enterprises have been taking place (Bloom et al. 2001; Anderson and Williams 2008; Sharma 2009). Unattached youth are defined as "those who are in the age group of 14–24 years, unemployed or outside the labour force, and not in school or in training" (HEART Trust/NTA 2009, p. 1). Many, but not all of these youth, fall into the category of at-risk youth where low levels of education and training, limited employment and lack of family and social support combine to create the risk of not being able to meet basic needs. The need for interventions for unattached youth has been highlighted in policy documents (National Centre for Youth Development 2004; Planning Institute of Jamaica 2011), given the increasing level of youth unemployment −30.8 % of youth in year 2010 (Planning Institute of Jamaica 2011). Agencies such as the HEART Trust/NTA have incorporated entrepreneurial training into their curriculum. However, there is still the need for increased focus on aspects such as training and education for entrepreneurs, particularly in building on innate "survival entrepreneurship" tendencies among the Jamaican population. Telecentres are positioned to provide support for these initiatives, through the services offered and their role in communities, and some have established training partnerships with HEART programmes.

Decision-making processes by telecentre staff and users are critical, and in evaluating the impact of telecentres, researchers need to understand these decision-making processes and who is making the decisions at the individual and group levels (Colle 2005; Whyte 2000). For example, Parkinson and Lauzon (2008) describe a case where decisions on telecentre activities were being made without adequate input from the community, resulting in limited telecentre success. How the decisions are made is also important, but there is limited research in this area. This suggests that telecentre implementations may benefit from the development of EDTMs that take into account day-to-day scenarios. It is therefore useful to recognize the potential role of social ties in encouraging the exploration of economic opportunities through telecentre usage (Bailey and Ngwenyama 2009).

Many telecentres are now recognizing this need and creating programmes that will attract "corner youth" to the centre (Bailey 2009). A newly opened telecentre in an urban community reported that some of their first customers were street boys (Sheil 2009). The manager of the funding institution which established this telecentre noted that "there is information available here so that people can perhaps set themselves up as an entrepreneur" (Sheil 2009, p. 10B). A young man, who had been a member of a gang overseas for nine years, highlighted the benefits of participating in computer skills training at a community centre (Thompson 2008). While there has been success with these initiatives, participants and organizers note the challenges faced in assisting others to use the resources offered. This group of unattached youth faces the challenges of survival associated with growing up on street corners and limited opportunities (Brown et al. 1995). An EDTM can assist in tailoring the programmes to meet the needs of community members.

2. **Deciding on the set of decision alternatives**

Given the development mandate of the telecentres, and the stakeholders' interest that had been expressed in developing entrepreneurial endeavours, the decision alternatives selected were as follows: Use the telecentre for entrepreneurial endeavours or do not use the telecentre for entrepreneurial endeavours.

3. **Conducting ethnographic** interviews

Ethnographic approaches can enable greater understanding of the impact of telecentre usage (Bailur 2007b; Dey et al. 2010) and decisions re-entrepreneurial endeavours (Gladwin et al. 1989). In this study, in-depth semi-structured interviews were conducted with telecentre staff and ethnographic interviews were conducted with users to learn about their experiences and identify the ideas that contribute to the decision-making processes of which telecentre programmes a user selects.

4. **Conducting participant observation**

Participant observation, along with the ethnographic interviews enabled a deeper understanding of the decision-making processes of the telecentre users.

The interviews and participant observation discussed in steps (4) and (5) provided an in-depth look at the usage of telecentres in the communities. Some of the findings are outlined here, and further details of the ethnographic interviews are described following step (6) to illustrate the creation of the individual decision trees.

Table 2 Entrepreneurial activities currently promoted by telecentre

Entrepreneurial programme	Activities
Business development	Entrepreneurial skills training; partnerships; community surveys
Community radio	Radio broadcasting
Community tourism	Community day tours; extended stays
Computer skills	Computer repair; word processing; data entry; website development
Creative writing	Short stories, poems
Music	Music production; disc jockey
Video	Video production
Visual arts	Computer graphics; digital photography

As described by stakeholders, entrepreneurial activities supported by various telecentres are outlined in Table 2. These activities have been developed in response to the perceived needs of the communities, the interests of community members and the potential demand for these services as a business. Telecentre staff, through consultation with community members, have identified what would attract the participation of various target groups in the community. One of the telecentre coordinators noted that "We decided to make the ICT programme attractive to get the youth off the corner". This involved creatively blending music and art programmes with functional and digital literacy programmes. Another telecentre coordinator noted the expectations of many participants of quick financial rewards from the training: "What they want to do is learn the things that can help them financially—like how to connect up a laptop to use with a sound system". Awareness of the possibilities which ICTs can be used for is increasing and there is much interest in the programmes being offered such as music and video production, graphic design and website development. Telecentre staff and users identified word-of-mouth as the main means by which community members learn about telecentre services. The majority of respondents in the survey heard about the telecentre from friends.

5. **Selecting the sample to use in the model**

The sample of telecentre users was selected based on those who were able to participate in ethnographic interviews on their decision whether to use the telecentre for entrepreneurial endeavours or not. This sample of 25 persons is used to develop the model.

6. **Deciding on the criteria to use in the model**

The reasons given were used to construct an initial model (Heemskerk 2002). Based on the interviews, a decision was made on criteria to use in the model, developing an individual decision tree for the first interview, and modifying or developing additional trees for subsequent interviews.

A brief description of the life situations of three telecentre users is presented below as they embark on entrepreneurial activities through the use of the telecentre. These ethnographic descriptions helped to inform the individual and composite decision tree models based on the issues they considered as they made their decisions.

The first telecentre user is a young man, seventeen years old, who lives in a volatile urban community. He has completed some years of secondary schooling, but was not trained for any specific occupation or trade. After leaving school, he became an apprentice to a mechanic, but the business has since closed and he is currently unemployed. He learned about the telecentre from a friend and uses it regularly to help with earning a living. He prints images on T-shirts and has been using the Internet to help with designing graphics. He also finds the telecentre staff helpful in assisting him with these designs on software programs available at the telecentre, as he states "Even if I can get to use a computer somewhere else, I come here because there is someone here who knows a lot about graphic design and helps me". His decision to use the telecentre for entrepreneurial endeavours was based on issues surrounding his search for an income-earning activity, his interest in graphic design—one of the visual arts programmes promoted by the

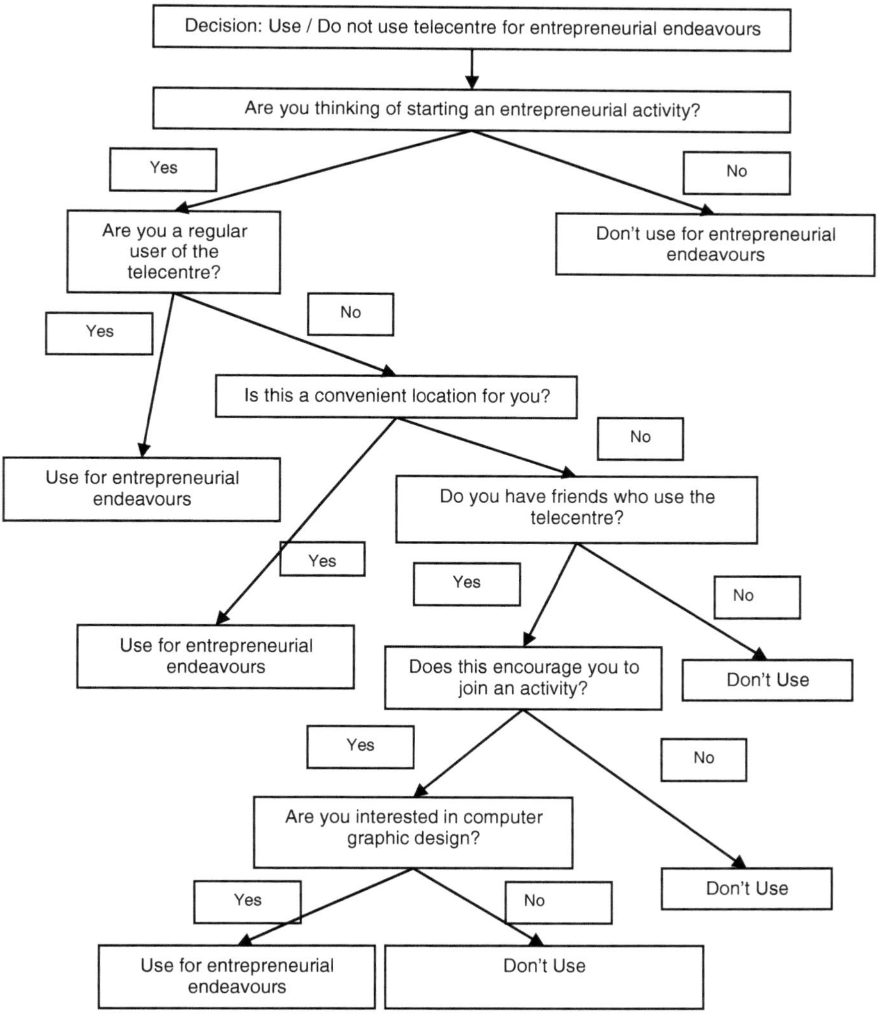

Fig. 1 Decision model for telecentre user 1

telecentre, and the recommendation of the telecentre by a friend. His decision to use the telecentre for entrepreneurial activity is depicted in Fig. 1.

Another telecentre user, a 27-year old single mother from a low-income urban community, left school at Grade 9. She is currently unemployed, but volunteers with a community-based organization. She is attending classes in food and beverage service at a national training academy. As an active member of a community-based organization, which is affiliated with the telecentre, she has been encouraged by telecentre staff and community members to start a business or seek employment with a company. She has decided to continue with her area of training, food and beverage, and spends her available free time at the telecentre searching for recipes, catering ideas, and sending resumes to restaurants and fast-food establishments. The process through which she decides to use the telecentre to assist with entrepreneurial ideas is illustrated in Fig. 2.

The third telecentre user is a 23-year old female from a rural community who has been trained in housekeeping for the hotel industry. She is unable to find a job and has enrolled in a website design course offered by the community telecentre. She is thinking of starting a business in the personal care

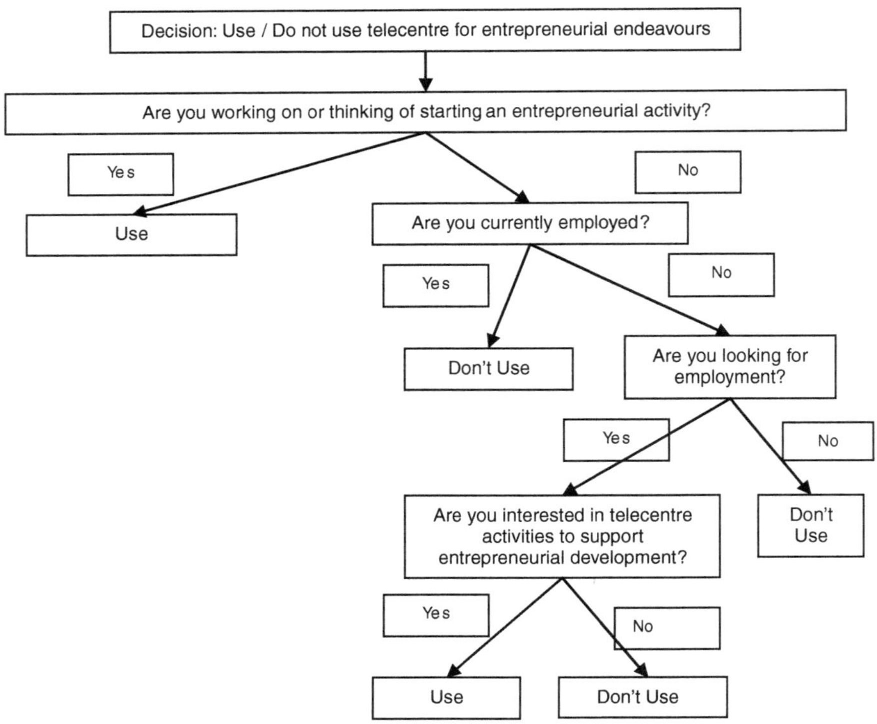

Fig. 2 Decision model for telecentre user 2

industry offering services such as manicures and pedicures. During the course, she has been working on designing the website for her planned business. She heard about the telecentre from friends and decided to use their services as it was close to her home and they partner with a national training agency to offer entrepreneurial courses. As she demonstrated her website she said "With this now I can promote the services, and I will also be doing business cards here". Her decision to use the telecentre for development of an entrepreneurial activity is depicted in Fig. 3.

7. Building the composite decision model for the group from the individual decision trees

The trees were then combined to create the composite tree. The model presented in Fig. 4 combines the individual decisions considered by telecentre users based on their life situations, in order to create an overall model of the process.

A sample of 25 telecentre users was used to test the model, which had an 88 % rate of accuracy of prediction. Three errors were identified between observed and predicted decisions. The model reflects the users' choices based on plans for entrepreneurial activities, familiarity with telecentre programmes, current employment

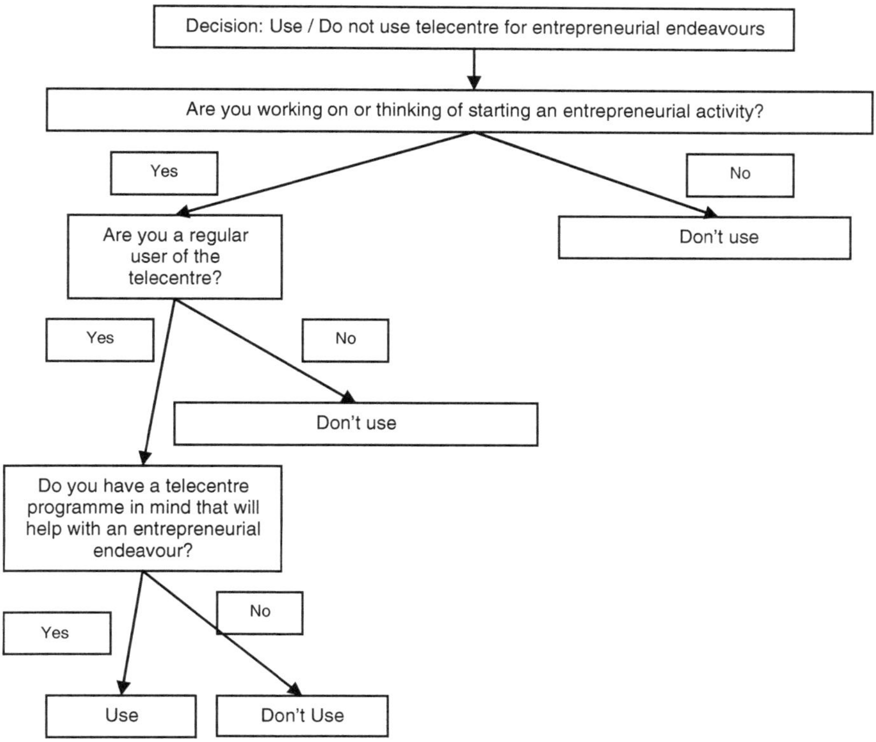

Fig. 3 Decision model for telecentre user 3

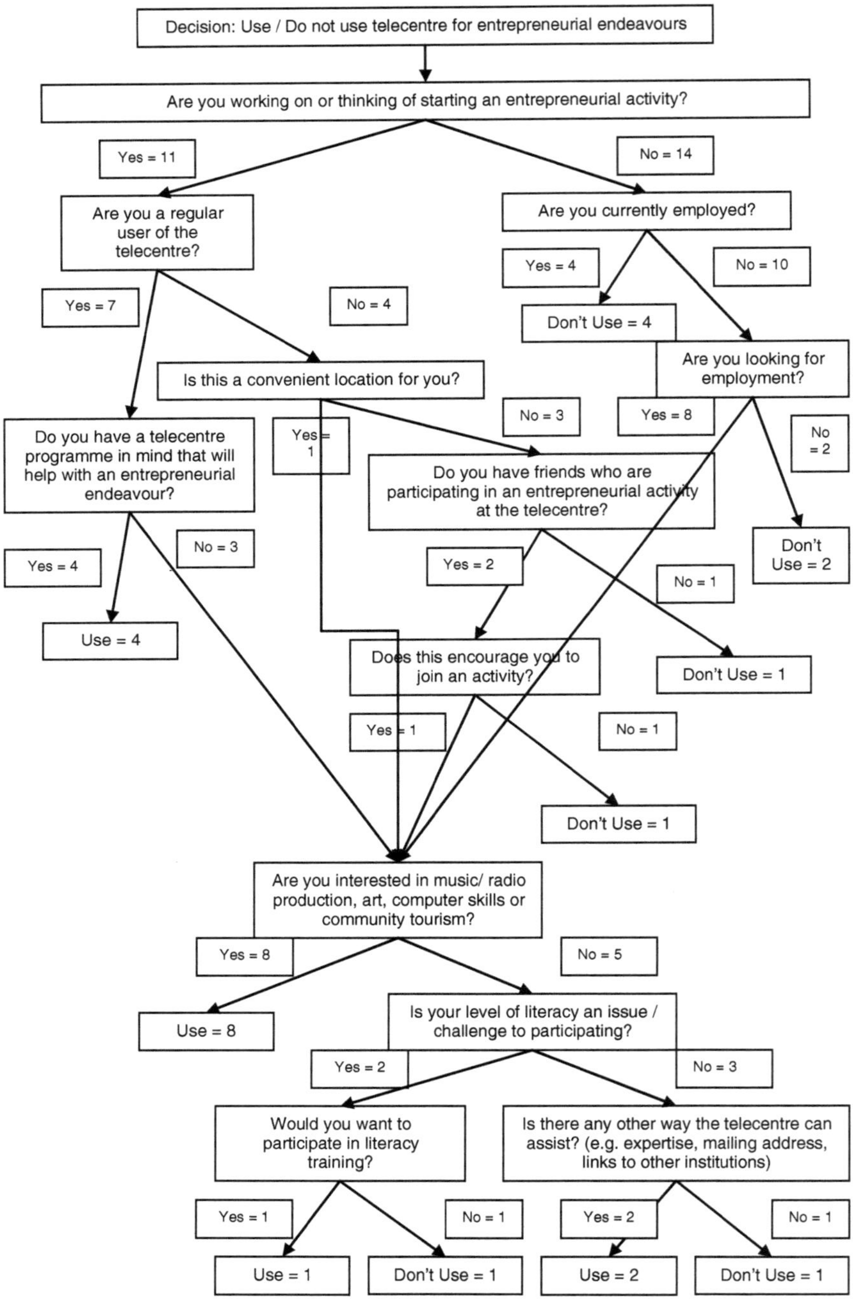

Fig. 4 EDTM of telecentre use for entrepreneurial endeavours (N = 25)

status, location of the telecentre, social ties, interest in programmes being offered by the telecentre and literacy levels.

The combination of decision processes depicted in the model shows the criteria that telecentre users may be employing in deciding whether to use the telecentre to assist with entrepreneurial endeavours. For example, a regular user of the telecentre, who may already have an entrepreneurial idea may be exposed to some of the particular ways that the telecentre can assist and decide to participate in a particular programme (4 users in the model). Another regular user, who may not be thinking of an entrepreneurial venture may be encouraged to use the available resources if the need arises, for example, due to exploring employment opportunities. Other persons who have a particular idea in mind may decide to use the telecentre services to assist, although the centre is not in the closest location, if they have friends who use the centre (1 user in the model). The areas of literacy and interest in employment were areas in which the model could be further enhanced. For example, the test of the model revealed 1 user who was not thinking of an entrepreneurial idea and was not looking for employment, but did use the telecentre to develop an entrepreneurial endeavour through recognition of an opportunity.

4 Discussion of Research Findings

There is increasing interest in factors which influence entrepreneurial behaviour and key elements which motivate nascent entrepreneurs (Sequeira et al. 2007). As these factors are investigated, it is important to examine the context in which they operate (Lundström and Stevenson 2005). Our study enabled the development of an EDTM which provides a means by which telecentre stakeholders can explain and predict the entrepreneurial decision-making process of telecentre users. Further, this EDTM provides insights and facilitates the development of a predictive model of entrepreneurial behaviour, discussed in Bailey and Ngwenyama (2013). Based on our empirical observations and analysis of decision criteria which emerged in our study, key concepts related to entrepreneurial behaviour by telecentre users were revealed. The decision criteria were grouped and mapped to the concepts of social ties, opportunity recognition and support from the telecentre as shown in Table 3.

Grounded in the empirical context, the concepts then form a predictive model which enables us to develop propositions which are grounded in the empirical observations. Bailey and Ngwenyama (2013) further discuss the components of the conceptual model and propositions are developed based on the empirical observations and supporting literature. The propositions represent relationships posited among the constructs, grounded in the empirical findings, that can be further developed into hypotheses and tested in future research.

Table 3 Mapping of decision criteria to key concepts

Decision criteria	Concept
Working on or thinking of starting entrepreneurial activity?	Opportunity recognition
Regular user of the telecentre?	Support from telecentre
Do you have a telecentre programme in mind that will help with an entrepreneurial endeavour?	Support from telecentre
Are you looking for employment?	Opportunity recognition
Do you have friends who are participating in an entrepreneurial activity at the telecentre?	Social ties
Does this encourage you to join an activity?	Social ties
Are you interested in music/radio production, art, computer skills or community tourism?	Opportunity recognition
Would you want to participate in literacy training?	Support from telecentre
Is there any other way the telecentre can assist? (e.g. expertise, mailing address, links to other institutions)	Support from telecentre

5 Conclusions and Implications for Future Research

As telecentres continue to work towards fulfilling their development mandate, there is a need for the identification of the process by which potential users make decisions about the use of the telecentre and the particular services they will need. Telecentre success depends on telecentre usage, which is influenced by a number of factors. NGOs and CBOs have been looking at strategies to increase participation in these programmes (Violence Prevention Alliance 2009). It has also been argued that in general, there should be greater exploration and facilitation of the possibilities of non-traditional forms of entrepreneurial activities (Boxill 2003). CBOs are well positioned to assist with entrepreneurial development (Ffrench 2008; Porter 1995). The success of community-based telecentres will be of benefit to developing countries, and as such, investigation of usage patterns and factors influencing decisions is useful.

EDTM provides a means of investigating and predicting the actual decision-making process of telecentre users which helps telecentre stakeholders to formulate strategies at many levels to encourage usage. It also assists with the identification of new initiatives, the change in usage demands over time and the evaluation of longer-term impacts. The findings discussed in this chapter and the ethnographic decision tree and predictive models that are presented facilitate knowledge sharing and feedback on ways to address the growing problem of unemployment, taking into account socio-economic impact and ensuring that the community members will actively participate in these approaches. While self-employment forms an important part of the economic activities of rural and urban Jamaicans (Anderson and Witter 1991), Anderson and Williams (2008) argue that there is a significant likelihood that youth who are self-employed may live in poverty and that programmes that promote self-employment should build linkages for employment opportunities within larger enterprises. Telecentres are among the community-based initiatives that try to place participants in their training programmes in internships during the programme and jobs on completion of the programme, while encouraging entrepreneurship by individuals or among groups in the community. One of the challenges they face, however, is migration of some community members following training. The telecentre staff and community organizers are proud of the progress and realize some of the benefits which can accrue from community members becoming successful outside of the community; however, they recognize that the community also needs resources within the community to facilitate development. Further, a supportive ecosystem may be useful in sustaining entrepreneurial activities embarked on by telecentre users.

Future research could test the EDTM with a wider group of telecentre users and community members and explore the concept of community-based entrepreneurship through the predictive conceptual model of entrepreneurial behaviour. The development of additional EDTM to predict decision-making processes for other scenarios related to telecentre usage will also enhance the development of theory in this area of ICT for development.

Acknowledgments We would like to thank the telecentre coordinators, staff, users and community members for their participation in this research.

References

Anderson P, Williams C (2008) The conversation between statistics and social policy: when we listen, when we don't. Dialogue for development lecture, Planning Institute of Jamaica, Kingston, Jamaica

Anderson P, Witter M (1991) Crisis, adjustment and social change: a case-study of Jamaica. United Nations Research Institute for Social Development and the Consortium Graduate School, University of the West Indies, Kingston, Jamaica

Bailey A (2009) Issues affecting the social sustainability of telecentres in developing contexts: a field study of sixteen telecentres in Jamaica. Electron J Inf Syst Developing Countries 36(4):1–18

Bailey A, Ngwenyama O (2013) Toward entrepreneurial behavior in underserved communities: an ethnographic decision tree model of telecenter usage. Information technology for development (ahead-of-print), pp 1–19

Bailey A, Ngwenyama O (2009) Social ties, literacy, location and the perception of economic opportunity: factors influencing telecentre success in a development context. In: Proceedings of the 42nd Hawaii international conference on system sciences, HICSS-42, January 5–8, Hilton Waikoloa Village, Island of Hawaii, pp 1–11

Bailur S (2007a) Using stakeholder theory to analyze telecenter projects. Inf Technol Int Dev 3(3):61–80

Bailur S (2007b) The complexities of community participation in ICT for development projects: the case of "our voices". In: Proceedings of 9th international conference on social implications of computers in developing countries

Blaikie NWH (2000) Designing social research: the logic of anticipation. Polity, Cambridge

Bloom DE, Mahal AS, King D, Mugione F, Henry-Lee A, Alleyne D, Castillo P, River Path Associates (2001) Jamaica: globalisation, liberalization and sustainable human development. UNCTAD/UNDP programme on globalisation, liberalization and sustainable human development

Boxill I (2003) Unearthing black entrepreneurship in the Caribbean: exploring the culture and MSE sectors. Equal Opportunities Int 22:32–45

Brown J, Newland A, Anderson P, Chevannes B (1995) Caribbean fatherhood: underresearched, misunderstood. Caribbean Child Development Centre and Department of Sociology and Social Work, University of the West Indies, Kingston, Jamaica

Cavaye ALM (1996) Case study research: a multi-faceted research approach for IS. Inf Syst J 6:227–242

Colle R (2000) Communication shops and telecenters in developing nations. Community Informatics: enabling communities with information and communications technologies, Idea Group Press, Hershey

Colle R (2005) Memo to telecenter planners. Electron J Inf Syst Developing Countries 21(1):1–13

Denzin NK, Lincoln YS (eds) (2000) Handbook of qualitative research, 2nd edn. Sage, Thousand Oaks

Dey BL, Newman DR, Prendergast R (2010) Ethnographic approach to user-centred evaluation of telecentres. Int J Innovation Digit Econ 1(3):22–39

Ellen D (2003) Telecentres and the provision of community based access to electronic information in everyday life in the UK. Inf Res 8(2):146

Ffrench S (2008) Funding entrepreneurship among the poor in Jamaica. Soc Econ Stud 57(2):119–148

Gladwin CH (1989) Ethnographic decision tree modeling. Sage, Thousand Oaks

Gladwin CH, Peterson JS, Mwale AC (2002) The quality of science in participatory research: a case study from Eastern Zambia. World Dev 30(4):523–543

Gladwin CH, Long BF, Babb EM, Beaulieu LJ, Mosely A, Mulkay D, Zimet DJ (1989) Rural entrepreneurship—one key to rural revitalization. Am J Agric Econ 71(5):1305–1314

Gomez R, Hunt P, Lamoureux E (1999) Telecentre evaluation and research: a global perspective. International Development Research Centre, Ottawa

Harris R, Kumar A, Balaji V (2003) Sustainable telecentres? Two cases from India. In: Krishna S, Madon S (eds) The digital challenge: information technology in the development context, Chap. 8, pp 124–135

HEART Trust/NTA (2009) Unattached youth in Jamaica, HEART Trust—National Training Agency, Kingston, Jamaica

Heemskerk M (2002) Livelihood decision-making and environmental degradation: small-scale gold mining in the Suriname Amazon. Soc Nat Resour 15(4):327–344

Hudson H (2001) Telecenter evaluation: issues and strategies. In: Latchem C, Walker D (eds) Telecenters: case studies and key issues. The Commonwealth of Learning, Vancouver

ICT4D Jamaica (2008) ICT4D interventions in Jamaica—a collection of case studies highlighting the use of ICTs in national development, vol I

Korsching PF, Allen JC (2004) Locality based entrepreneurship: a strategy for community economic vitality. Community Dev J 39:385–400

Kuriyan R, Ray I (2009) Outsourcing the state? Public-private partnerships and information technologies in India. World Dev 37:10

Kvasny L, Keil M (2006) The challenges of redressing the digital divide: a tale of two US cities. Inf Syst J 16(1):23–53

Lundström A, Stevenson L (2005) Entrepreneurship policy: theory and practice. Kluwer Academic Publishers, Boston

Madon S (2005) Governance lessons from the experience of telecentres in Kerala. Eur J Inf Syst 14(4):401–416

Madon S, Reinhard N, Roode D, Walsham G (2009) Digital inclusion projects in developing countries: processes of institutionalisation. Special issue: Development and the promise of technological change. Inf Technol Dev 15(2):95–107

Nandhakumar J, Jones M (1997) Too close for comfort? Distance and engagement in interpretive information systems research. Inf Syst J 7:109–131

National Centre for Youth Development (2004) National Youth Policy, Ministry of Education, Youth and Culture, Kingston, Jamaica

Ngwenyama O, Andoh-Baidoo F, Bollou F, Morawczynski O (2006) Is there a relationship between ICT, health, education and development? An empirical analysis of five West African Countries from 1997–2003. Electron J Inf Syst Developing Countries 23(5):1–11
Osei-Bryson KM (2004) Evaluation of decision trees: a multi-criteria approach. Comput Oper Res 31(11):1933–1945
Parkinson S, Lauzon A (2008) The impact of the internet on local social equity: a study of a telecenter in aguablanca, Colombia. Inf Technol Int Dev 4(3):21–38
Planning Institute of Jamaica (2011) Economic and social survey of Jamaica: 2010. Planning Institute of Jamaica, Kingston
Porter ME (1995) The competitive advantage of the inner city. Harvard Bus Rev 55–71 (May–June)
Ryan G, Bernard H (2000) Data management and analysis methods. In: Denzin N, Lincoln Y (eds) Handbook of qualitative research, 2d edn. Sage, Thousand Oaks, pp 769–802
Ryan G, Bernard H (2006) Testing an ethnographic decision tree model on a national sample: recycling beverage cans. Hum Organ 65(1):103–114
Sequeira J, Mueller SL, Mcgee JE (2007) The influence of social ties and self-efficacy in forming entrepreneurial intentions and motivating nascent behaviour. J Dev Entrepreneurship 12(3):275–293
Sey A, Fellows M (2009) Literature review on the impact of public access to information and communication technologies. Working paper no. 6, Seattle, WA, TASCHA (CIS)
Sharma A (2009) Small businesses a driving force for the economy. The Sunday Gleaner, Kingston, Jamaica, p C8
Sheil R (2009) 'The source' reaches Maverley. The Jamaica Observer, Kingston, Jamaica, p 10B
Thompson S (2008) Gangster turns student—former bad boy tells IMF boss how far he has come. The Gleaner, Kingston, Jamaica, p A2
Violence Prevention Alliance (2009) The peace guardian, violence prevention alliance newsletter, vol 1(6). Kingston, Jamaica January
Walsham G, Robey D, Sahay S (2007) Foreword: special issue on is in developing countries. MIS Quart 31(2):317–326
Whyte A (2000) Assessing community telecentres: guidelines for researchers. International Development Research Centre (IDRC), Canada

Chapter 7
Using Association Rules Mining to Facilitate Qualitative Data Analysis in Theory Building

Yan Li, Manoj Thomas and Kweku-Muata Osei-Bryson

The richness captured in qualitative data is a key strength of the qualitative approach to theory building. However, given the nature of qualitative data, it is typically not apparent to qualitative researchers as to how quantitative techniques could be used to facilitate the identification of strong relationships between concepts that are embedded in the data. This typically leads to the formulation of theoretical propositions that are often rich in detail, yet lacking in simplicity. In addition, the researcher faces the daunting task of developing persuasive arguments to justify the findings. This chapter proposes a systematic procedure toward qualitative data analysis to facilitate developing propositions in theory building. Specifically, we demonstrate how researchers can take advantage of quantitative data analysis techniques such as association rules (AR) mining to identify strong concept relationships from qualitative data. The proposed procedure is illustrated using a case study in the public health domain.

1 Introduction

Theories can be broadly defined as coherent descriptions or explanations of reality (Gioia and Pitre 1990). According to Dubin (1978, p.26), *A theory tries to make sense out of the observable world by ordering the relationships among elements*

Y. Li (✉) · M. Thomas · K.-M. Osei-Bryson
Department of Information Systems, Virginia Commonwealth University,
301 W. Main Street, Richmond VA 23284, USA
e-mail: LIY26@VCU.Edu

M. Thomas
e-mail: mthomas@VCU.Edu

K.-M Osei-Bryson
e-mail: KMOsei@VCU.Edu

K.-M. Osei-Bryson and O. Ngwenyama (eds.), *Advances in Research Methods for Information Systems Research*, Integrated Series in Information Systems 34,
DOI: 10.1007/978-1-4614-9463-8_7,

that constitute the theorist's focus of attention. Theory building is a process by which the theoretical presentation is generated, tested, and/or refined. Gioia and Pitre (1990) identify four steps toward theory building: *opening work, data collection, analysis*, and *theory building*. *Opening work* involves research topic identification and research design. Based on the research design, *data collection* involves gathering data that are relevant to the research using techniques such as surveys, archival data, interviews, observations, etc. *Analysis* can take different forms based on the nature of research design and data collected. Traditional positivist approaches use deductive reasoning and causal analysis to evaluate the significance of the data, while interpretive approaches advocate inductive reasoning to identify emergent concepts and relationships. The last step is *theory building* where the findings related to the phenomenon of interest are summarized as a theoretical representation.

Limitations of an existing theory are exposed when a particular phenomenon arises that is not explainable by the theory. Under such circumstance, positivist approaches often fall short as they rely on verification or falsification of the hypothesis and consequently are limited to incremental revision or extension of the original theory. On the other hand, theory building based on qualitative methods centers directly on the juxtaposition of contradictory evidence to provide novel insights (Eisenhardt 1989), thereby addressing the above shortcomings of quantitative approaches. The strength of qualitative methods in theory development and refinements relies in the rich knowledge captured in the qualitative data. It helps to develop theories that are relevant, rich, and dynamic in their explanations of social processes (Fine and Elsbach 2000).

Although the richness of qualitative data is invaluable, it can also pose a challenge to the theory development process. In the quest to explain the data, a qualitative researcher can easily go in the direction of making theoretical propositions that are rich in detail, yet lacking in simplicity, a key dimension for good theory (Weick 1979). The nature of the qualitative data precludes the researcher from using quantitative techniques such as statistical testing to identify strong relationships between concepts that are identified through the coding process. In addition, the researcher faces the daunting task of developing persuasive arguments to justify the findings. A systematic approach toward qualitative data analysis is therefore compelled to facilitate developing propositions, and establishing the strength and consistency of the findings. To the best of our knowledge, this aspect of analysis has not been addressed. For example, grounded theory (Glaser and Strauss 1967), a widely used theory building methodology in social science, introduces constant comparisons as its data analysis strategy to explain patterns in qualitative data. However, grounded theory building is a descriptive process rather than a prescriptive one. It is left to the researcher to justify when theoretical saturation is achieved (i.e., further refinement of the concepts and their relationships add little to that which has already surfaced). Yin (2003) has recommended several qualitative data analysis tactics (i.e., pattern matching, explanation building, addressing rival explanations, and logic models) to test validity and reliability of the qualitative research design. However, these tactics only test internal validity for explanatory or causal studies, whereas studies in theory building are exploratory in nature. Miles and Huberman (1994) have presented three types of qualitative data analysis activities (i.e., data reduction or coding, data display in matrix or graphs, and conclusions drawing and verification). Their focus is on managing and representing qualitative data without losing their meanings through intensive coding. Eisenhardt (1989) has recommended common qualitative data analysis techniques such as within-case analysis and cross-case pattern search when building theory from case study research. None of these aforementioned data analysis techniques assist the systematic identification and justification of concept relationships.

The objective of this chapter is to demonstrate how researchers can take advantage of quantitative data analysis techniques such as association rules (AR) mining to identify strong concept relationships using qualitative data. The underlying philosophical differences between qualitative and quantitative researches do not prevent the combination of the two. There are several contexts (Denzin and Lincoln 1994; Gioia and Pitre 1990; Jick 1979) where both have been used in conjunction during theory building. However, these contexts focus on either research design or data collection, where the data analysis part receives little attention. The rest of this chapter is organized as follows. We first provide an overview of AR induction, followed by a description of the proposed procedure to analyze qualitative data using quantitative techniques. The procedure is illustrated using a case study in the public health domain. Finally, we conclude by stating the contributions of this study.

2 Overview of Associate Rules Induction

AR mining is a popular pattern discovery method in knowledge discovery and data mining (KDDM). It was first introduced by Agrawal et al. (1993) to mine large transactional databases. A transactional dataset for AR analysis can be defined in the following general terms. Let $I = \{i_1, i_2, \ldots, i_m)$ be a set of distinct items, and $T = \{t_1, t_2, \ldots, t_k)$ be a set of k subsets of I. Each t_i is a transaction such that $t_i \subseteq I$. For example, in market basket analysis, each basket is a transaction that contains the set of items purchased from one register transaction, and the set I consists of the items stocked by a retail outlet.

The objective of AR mining is to find items that imply the presence of other items in the same transaction. It can be expressed as $A \Rightarrow B$ (e.g., bread => peanut butter and jelly), where A and B are sets of items in a given transaction t_i, and $A \Rightarrow B$ meets both the minimal *support* and minimal *confidence* constraints. *Support* specifies the probability that a transaction t_i contains both items A and B. *Confidence* specifies the conditional support, given that the transaction already contains A. It should be noted that an AR does not always imply causation. Both *support* and *confidence* constraints are probability-based measures.

Table 1 Some interestingness measures for AR

Measure	Formula
Support	$P(AB) = \frac{n(AB)}{N}$
Confidence	$P(B\|A) = \frac{P(AB)}{P(A)}$
Coverage	$P(A) = \frac{n(A)}{N}$
Lift	$\frac{P(B/A)}{P(B)}$
Collective strength	$\frac{P(AB)+P(\neg B\|\neg A)}{P(A)P(B)+P(\neg A)\times P(\neg B)} \times \frac{1-P(A)P(B)-P(\neg A)\times P(\neg B)}{1-P(AB)-P(\neg B\|\neg A)}$
Expected confidence	$P(B)$
Reliability	$P(B/A) - P(B)$

The advantage of AR mining lies in finding all possible associations between relevant factors and presenting results in a simple and understandable manner. It has been applied to uncover interesting patterns in different application areas such as market basket analysis (Agrawal and Srikant 1994), Web mining (Liu et al. 2004), safety science (Montella 2011), medical records analysis (Chang 2007), and questionnaire analysis (Chen and Weng 2009). While AR mining has its key strength in its *understandability* and *completeness* (Liu et al. 1999), not all ARs are interesting. The candidate AR set often contains a large number of associations, making it difficult, and sometimes impossible to comprehend. Additional forms of rule interestingness measures have been developed to evaluate and select ARs based on their potential interestingness to the user (Geng and Hamilton 2006). Examples of these interestingness measures include *coverage* (Piatetsky-Shapiro and Fayyad 1991), *lift* (Brin et al. 1997), *collective strength* (Aggarwal and Yu 2001), and *reliability* (Ahmed et al. 2000). The *reliability* measure proposed by Ahmed et al. (2000) measures the difference between *confidence* and *expected confidence* of an AR, which is the effect of A on the probability of B. Because the *reliability* measure is a probability, it can be used in classical hypothesis testing. Table 1 shows the mathematical representation of the objective measures, where $n(A)$ denotes the number of transactions that contain A and $n(AB)$ denotes the number of transactions that contain both A and B, N denotes the total number of transactions, $P(A)$ denotes the probability of A, $P(\neg A)$ denotes probability of *not* A, and $P(B/A)$ denotes the conditional probability of B.

3 Description of Proposed Procedure

As mentioned earlier, the objective of this chapter is to demonstrate how data mining techniques such as AR can be applied to discover strong associations among concepts identified from qualitative data. The qualitative data need to be first coded into a transactional dataset, where each transaction (ti) is a set of concepts that are identified from each qualitative data point (such as an interview statement). The set I consists of a priori concepts identified before coding. AR can then be used to discover the underlying associations among concepts. The AR analysis consists of the following four steps:

1. **Data Preparation**: Preprocess the coded data into a transactional dataset.
2. **AR Discovery**: Use AR mining tool to discover candidate AR sets based on classic AR approach comprising of minimal confidence and minimal support.
3. **AR Pruning**: Prune the candidate ARs using selected *Interestingness measures*.
4. **Proposition Development**: Use the pruned set of ARs to guide the development of the theoretical propositions that describe relationships between the concepts.

In Sect. 4, we demonstrate the procedure using a case study from the public health domain. Related theories applicable to the case study are summarized, followed by the concept identification and coding steps, and the AR mining procedure for relationship discovery.

4 Illustration of the Proposed Procedure

4.1 Case Study Background

The case study is a cross-disciplinary research in the public health domain conducted to understand the health-seeking behavior of women in low-income low-resource communities of developing countries. To develop intervention programs aimed at reducing the burden of ill health in low-income communities (Huda 2006), the first step is to trace the health behavior patterns prevalent among its members. In a region where the socioeconomic welfare is low and the people live in a 'below poverty line' situation, an attempt to explain the high incidence of specific illness such as reproductive tract infections (RTI) requires identifying key associations among the relevant contextual factors. AR mining fits the explorative nature of the study by providing a means to analyze the qualitative data using predetermined objective measures.

The study was carried out in a marginalized community comprising of 510 families, of which 79 % of the men are employed in the fisheries sector. Due to the seasonal nature of employment and small scale of operations, men work a maximum of 5 months in a year. The women (approximately 200) mostly work as contract laborers in fish-processing jobs. The socioeconomic welfare of the respondents in the region is low (the average income ranges from \$200–\$450 perannum). Although the Panchayat (village) census lists all the houses in the ward with male household heads, primary data reveal that women are the virtual heads of the family. Thus, the burden of running the family rests on the shoulders of the women. To a large extent, unequal opportunities to act account for the prevalence of communicable and non-communicable diseases that go unchecked. The community is an exemplar of a subsistence economy. Open access mode of resource use, technological dualism, lack of educational attainment and class identity, and the absence of strong sociopolitical movements from within the community are push factors that retain them as a marginalized group (KDR 2008). Established theories from the public health domain served to set the focus for the problem analysis. The theories that were used in concept identification are discussed in Sect. 4.1.1

4.1.1 Related Health Behavior Theories

Many models have been proposed to explain the notion of health behavior in individuals. Munro et al. (2007) conducted a detailed literature review covering scholarly databases, electronic libraries, and citations to identify theories and models developed for the domain of 'health and behavior.' The study identifies nine prominent theories—behavioral learning theory (BLT), health belief model (HBM), social cognitive theory (SCT), theory of reasoned action (TRA), protection motivation theory (PMT), theory of planned behavior (TPB), information–motivation–behavioral (IMB) *skills model,* self-regulatory theory (SRT), and *the* transtheoretical model (TTM). The models have been applied in various settings ranging from developing interventions for cervical cancer prevention among Latina immigrants (Scarinci et al. 2011) to identifying the variables that can influence a smoker's motivation to quit (Norman et al. 1999).

Among them, HBM (Rosenstock et al. 1994) is perhaps the most widely used theory to explain health-related behavior (Urrutia 2009). It takes a cognitive perspective to explain and predict preventive health behavior (Hayden 2008). The six main constructs in HBM are the following: *perception of susceptibility to a disease or condition, perceived severity, perceived benefits of care, cues to action, self-efficacy,* and *barriers to preventive behavior* (Strecher and Rosenstock 1997). The model suggests that the *perceived seriousness* and *susceptibility to a disease* influences an individual's perception of the threat of disease. It posits that the likelihood of engaging in a recommended health behavior is based on an assessment of the benefits of a health-modifying action to the barriers in place (Munro et al. 2007; Urrutia 2009). *Self-efficacy* is the person's conviction to produce behavioral outcome (Hayden 2008; Scarinci et al. 2011). *Cues* are the bodily (e.g., identified symptoms) or environmental events (e.g., information sourced by media, reminders, incentives, or information imparted by peers and family members) that prompt an individual to adopt health-modifying actions (Hayden 2008; Strecher and Rosenstock 1997). The main components of the HBM and the key variables under each category are shown in Fig. 1.

All behavior-predicting models in the public health domain are conclusively drawn from empirical studies of population in developed nations. The models were developed based on data collected from the strata of society that have easy access to preventive and diagnostic services. Even the popular HBM finds greater applicability to middle-class groups than lower-status groups (Rosenstock 2005).

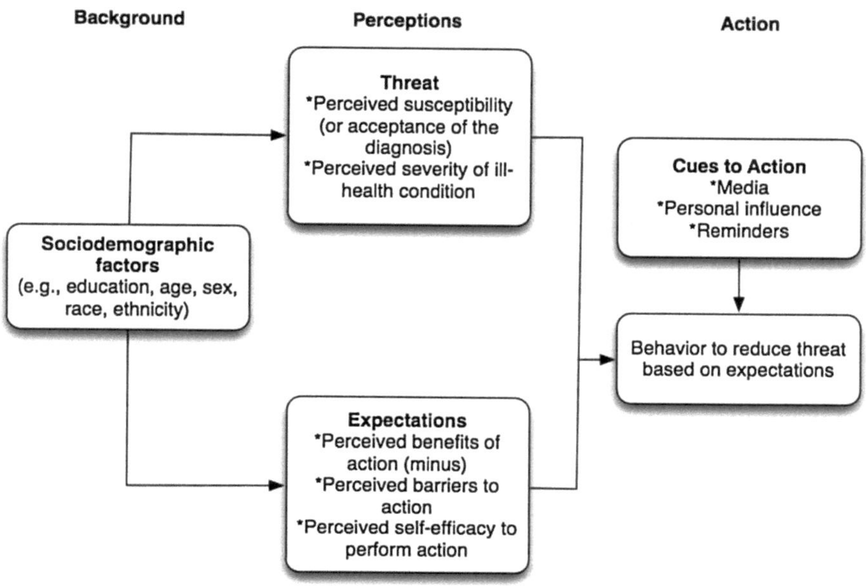

Fig. 1 The health belief model (adapted from Rosenstock et al. 1994)

The existing models thus fail to predict the health behavior among the underserved communities. It highlights the need for refining the theoretical conceptualizations to account for those living in the very low-income and low-resource societies.

4.1.2 Data Collection and Coding

Data were collected over a period of eight months using qualitative methods (Glaser and Strauss 1967) that involved face-to-face interviews, focus group discussions, and iterative follow-up meetings. All interviews were anonymized and transcribed before analysis. Three subject experts systematically analyzed statements from 8 subjects and 7 focus groups. The analysis followed the coding technique suggested by Hammersley and Atkinson (2007) and Denzin (1997). A total of 158 statements were evaluated in this manner.

The first step is to apply content analysis to identify the key concepts from the narrative data. Based on the HBM, the interview statements were coded across two dimensions consisting of 10 concepts—four subjective constructs and six behavioral factors. The subjective constructs are the *belief, desire, intention, and likelihood of action*. The factors influencing behavioral outcomes are *susceptibility, severity, benefits, barriers, cues to action*, and *self-efficacy*. In addition, the descriptive properties of the environment (socioeconomic and cultural structure) that shape the likelihood of adopting a health behavior change were also coded by the researchers.

4.2 Application of the Procedure

Step 1: Data Preparation

AR mining requires preprocessing data into a transactional dataset that includes multiple transactional items in the same transaction. In our case, the transaction is the equivalent of one coded statement by one researcher, and the transactional item is the equivalent of the individual concept in each statement.

To uniquely represent each coded statement, an ID is assigned to the coding result in the format of S_x–R_y–Z, where S_x is the subject ID, R_y is the research ID, and Z is the statement ID. The next step is to transform the coded items by assigning a '1' for the concept identified in S_x–R_y–Z, and '0' otherwise. The final dataset includes a total of 474 transactions and 1,331 transactional items. Table 2 shows the summary of transaction counts.

Step 2: AR Discovery

For this case study, we use the SAS Enterprise Mining 9.3 as the AR mining tool. The *association* node is used to extract candidate ARs with parameter settings of 10 % minimal confidence, 5 % minimal support, and leaving all other settings as default. Figure 2 displays the rule matrix for all candidate ARs before pruning. The candidate AR set includes 91 2-item or 3-item ARs, with the *confidence* between [15.63, 69.32] and *lift* between [0.25, 4.32].

Step 3: AR Pruning

Sub-Step 3a: Pruning using Reliability

To determine the ARs that are significant, the reliability score and t statistic for all candidate ARs are first calculated. The reliability is calculated using the confidence (C) and expected confidence (EC) from the output table, where *reliability* = (C − EC). Based on this, the following population proportion hypothesis testing is performed:

H_0 The difference between the confidence of the AR and the *expected confidence* of the AR is not statistically significant

H_1 The difference between the *confidence* of the AR and the *expected confidence* of the AR is statistically significant

A one-tail t test is then performed, and forty-nine (49) ARs were found to be statistically significant at significance level $\alpha = 0.01$ (supporting H_1), which means that those ARs have a greater confidence level than expected. The remaining 42 ARs were pruned.

Table 2 Transaction count summary

Concept	Count	Concept	Count
Belief	226	Susceptibility	89
Desire	136	Severity	133
Intention	123	Benefits	100
Likelihood	64	Barriers	258
		Cues	87
		Efficacy	116

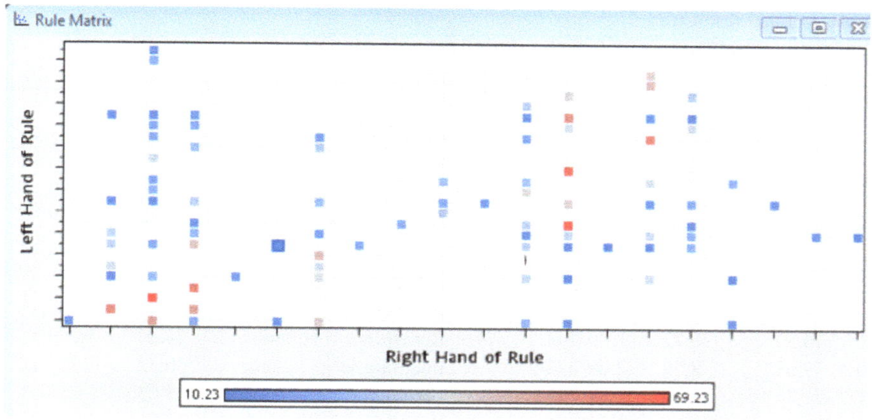

Fig. 2 Candidate rule matrix

Sub-Step 3b: Pruning using Understandability

The example presented is an exploratory case study of a problem domain that is inherently complex. To comprehend the interactions between the concepts, the *understandability* of the AR is a fitting interestingness measure. The number of items in a given AR has been shown to be such a measure (Freitas 1999), where ARs with fewer antecedents (fewer items in A) are considered to be easier to understand. Similarly, Geng and Hamilton (2006) consider the conciseness of an AR as an important perspective in rule interestingness measures. A concise AR contains relatively fewer items and is thus easier to assimilate. Hence, all ARs that have more than two items were pruned. This leaves a total of 22 ARs.

Sub-Step 3c: Pruning using Lift

The premise of this study is to find associated concepts that are departing from independence and positively correlated. *Lift* is a measure of departure from independence (Brin et al. 1997). A lift value greater than 1 means that *A* and *B* appear more frequently together than expected under independence, and vice versa. Thus, all ARs that have lift value less than or equal to 1 are pruned. This results in 20 ARs, which is shown in Table 3.

Research experts familiar with the research setting will be particularly interested in concepts with strong relationships. The ARs presented in Table 3 are organized in pairs, where each pair contains the same rule items, but the antecedent and the consequent are switched. Thus, each pair can be expressed as (*A* => *B* and *B* => *A*). As mentioned previously, an AR does not imply causality. This indicates a total of ten pairs of strong ARs that will require careful interpretation.

For better understandability of the results, we only consider two-item ARs in the illustrative example presented above. It is possible that AR mining may not have identified certain expected associations. Furthermore, the illustrative example

Table 3 Final result ARs

AR ID	Confidence (%)	Support (%)	Lift	Transaction count	AR	Reliability (%)	*T*
1	68.06	10.43	1.72	49	Susceptibility ==> belief	28.48	246.35*
2	26.34	10.43	1.72	49	Belief ==> susceptibility	11.02	181.45*
3	60	14.68	1.48	69	Desire ==> barrier	19.36	194.38*
4	36.13	14.68	1.48	69	Barrier ==> desire	11.66	170.14*
5	56.76	8.94	1.43	42	Cues ==> belief	17.18	137.60*
6	22.58	8.94	1.43	42	Belief ==> cues	6.84	102.48*
7	53.13	7.23	2.4	34	Likelihood ==> efficacy	31	339.07*
8	32.69	7.23	2.4	34	Efficacy ==> likelihood	19.08	280.15*
9	46.59	8.72	2.09	41	Benefit ==> intention	24.25	289.51*
10	39.05	8.72	2.09	41	Intention ==> benefit	20.32	271.14*
11	43.75	5.96	2.34	28	Likelihood ==> benefit	25.03	275.91*
12	31.82	5.96	2.34	28	Benefit ==> likelihood	18.2	242.58*
13	43.75	10.43	1.11	49	Severity ==> belief	4.18	36.12*
14	26.34	10.43	1.11	49	Belief ==> severity	2.51	31.47*
15	33.33	7.45	1.4	35	Intention ==> severity	9.5	100.52*
16	31.25	7.45	1.4	35	Severity ==> intention	8.91	98.27*
17	31.82	5.96	1.34	28	Benefit ==> severity	7.99	75.57*
18	25	5.96	1.34	28	Severity ==> benefit	6.28	69.20*
19	24.04	5.32	1.08	25	Efficacy ==> intention	1.7	15.83*
20	23.81	5.32	1.08	25	Intention ==> efficacy	1.68	15.78*

* $p < 0.01$

presented here only uses objective interestingness measures (confidence, lift, simplicity, and reliability) in the AR pruning step. This may not identify the most important association patterns to the researcher (McGarry 2005). Including subjective interestingness measures (such as actionable, unexpected, and novelty) in the evaluation can provide a deeper understanding of the problem domain.

Step 4: Proposition Development

The output of AR pruning (Step 3) results in a set of strong ARs, each of which can be a candidate proposition. The candidate proposition may substantiate into a theoretical proposition only if a persuasive explanation can be provided by the researcher. This assessment is achieved by the researcher's constant reflection on the phenomenon under the study, the evidence gathered from the case, and the existing literature and theory. The candidate propositions may quite likely be expected relationships or unexpected relationships. Hence, the researcher needs to first revisit the case evidence to determine why expected candidate propositions hold. For example, the AR pair, benefit and intention, forms two candidate propositions (AR9 and AR10 from Table 3). Reflecting on the data acquired from the fieldwork, it is observed that the women in the marginalized communities tend to pursue a health-improving behavior when the benefits of such action (e.g., benefits for the immediate family members assured by one's health as a motivating factor to seek medical care) are seen to reduce the disease

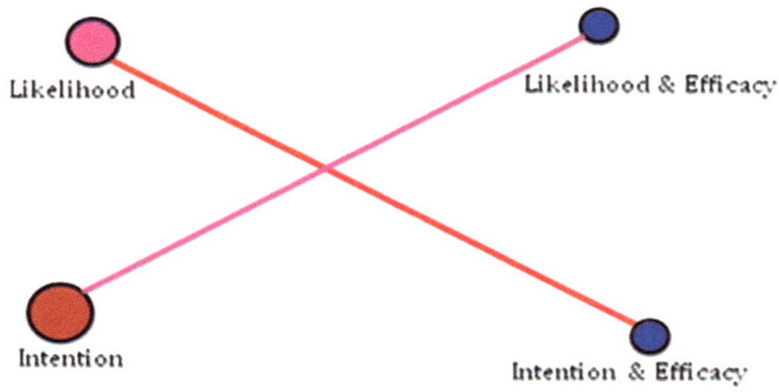

Fig. 3 SAS linkage graph result for intention and likelihood of action

threat. Hence, AR9 is suitable for further proposition development, and AR10 is eliminated. The higher reliability of AR9 (reliability = 24.25) provides additional support for this selection. In case of unexpected relationships, the case evidence can provide the researcher with ample opportunity and flexibility to gain new insights.

Another essential step is to compare the candidate propositions against existing literature to determine whether they are similar to, or contradict previous studies. Understanding the theoretical reasons as to why specific relationships exist demonstrates the internal validity of the findings. Where conflicting relationships are found, the researcher has to seek additional evidence to discern the plausible reasons, and reconcile the findings. For example, folk psychology (Malle and Knobe 1997) suggests a significant association between intention and likelihood of action, though this association did not emerge as a strong AR from the analysis. This motivates the research to go back to the data and perform additional AR mining to include three or more items in ARs. For instance, the SAS linkage graph shows the possibility of association between intention and likelihood with efficacy as an influencing factor (Fig. 3). Only after these iterative steps are complete, a set of theoretical propositions can be expounded. The set of theoretical propositions can then be summarized into a logical, systematic explanatory schema for future theory testing (Glaser and Strauss 1967).

5 Conclusion

This chapter demonstrates the use of AR to determine the associations among concepts identified via content analysis of qualitative data. Researchers working with voluminous qualitative data often struggle to find an even ground between the richness of the observations and the simplification of the findings. AR provides a quantitative gauge to assess the important construct relationships during the formative phase of theory building. From a methodological viewpoint, the illustrative example shows a novel application of AR mining technique for theory building using qualitative data.

References

Aggarwal CC, Yu PS (2001) Mining associations with the collective strength approach. IEEE Trans Knowl Data Eng 13(6):863–873
Agrawal R, Imieli T, Swami A (1993) Mining association rules between sets of items in large databases. SIGMOD Rec 22(2):207–216
Agrawal R, Srikant R (1994) Fast algorithms for mining association rules. In: Proceedings of the 1994 international conference on very large databases, USA
Ahmed KM, El-Makky NM, Taha Y (2000) A note on "beyond market baskets: generalizing association rules to correlations". SIGKDD Explor Newsl 1(2):46–48
Brin S, Motwani R, Silverstein C (1997) Beyond market baskets: generalizing association rules to correlations. Paper presented at the proceedings of the 1997 ACM SIGMOD international conference on Management of data
Chang C-L (2007) A study of applying data mining to early intervention for developmentally-delayed children. Expert Syst Appl 33(2):407–412
Chen Y-L, Weng C-H (2009) Mining fuzzy association rules from questionnaire data. Knowl-Based Syst 22(1):46–56
Denzin N, Lincoln Y (1994) Handbook of qualitative research. Sage, Thousand Oaks
Denzin NK (1997) Interpretive ethnography—ethnographic practices for the 21st century. Sage Publications, Thousand Oaks
Dubin R (1978) Theory building. Free Press, New York
Eisenhardt KM (1989) Building theories from case study research. Acad Manag Rev 14(4):532–550
Fine GA, Elsbach KD (2000) Ethnography and experiment in social psychological theory building: tactics for integrating qualitative field data with quantitative lab data. J Exp Soc Psychol 36(1):51–76
Freitas AA (1999) On rule interestingness measures. Knowl-Based Syst 12(5–6):309–315
Geng L, Hamilton HJ (2006) Interestingness measures for data mining: A survey. ACM Comput Surv 38(3):9
Gioia DA, Pitre E (1990) Multiparadigm perspectives on theory building. Acad Manag Rev 15(4):584–602
Glaser BG, Strauss A (1967) The discovery of grounded theory: strategies for qualitative research. Aldine Publishing, Chicago
Hammersley M, Atkinson P (2007) Ethnography: principles in practice (3 edn). Routledge, Taylor and Francis Group, London
Hayden JA (2008) Health belief model introduction to health behavior theory, 1st edn. Jones and Bartlett Publishers, Burlington
Huda Z (2006) Study design for the measurement of gynaecological morbidities and health seeking behaviour. In: Jeejebhoy S, Koenig M, Elias C (eds) Investigating reproductive tract infections and other gynaecological disorders—a multi-disciplinary research approach. Cambridge University Press, Cambridge
Jick TD (1979) Mixing qualitative and quantitative methods: triangulation in action. Adm Sci Q 24(4):602–611
KDR (2008) Kerala Development Report. Academic Foundation, New Delhi
Liu B, Grossman R, Zhai Y (2004) Mining web pages for data records. IEEE Intell Syst 19(6):49–55
Liu B, Hsu W, Ma Y (1999) Pruning and summarizing the discovered associations. Paper presented at the proceedings of the fifth ACM SIGKDD international conference on Knowledge discovery and data mining
Malle BF, Knobe J (1997) The Folk concept of intentionality. J Exp Soc Psychol 33:101–121
McGarry K (2005) A survey of interestingness measures for knowledge discovery. Knowl Eng Rev 20(01):39–61
Miles MB, Huberman AM (1994) Qualitative data analysis. Sage, Thousand Oaks

Montella A (2011) Identifying crash contributory factors at urban roundabouts and using association rules to explore their relationships to different crash types. Accid Anal Prev 43(4):1451–1463

Munro S, Lewin S, Swart T, Volmink J (2007) A review of health behaviour theories: how useful are these for developing interventions to promote long-term medication adherence for TB and HIV/AIDS? BMC Public Health 7:104

Norman P, Conner M, Bell R (1999) The theory of planned behavior and smoking cessation. Health Psychol 18(1):89–94

Piatetsky-Shapiro G, Fayyad U (eds) (1991) Knowledge discovery in databases. MIT AAAI Press, California

Rosenstock IM (2005) Why people use health services. Milbank Fund Q 83(4):

Rosenstock IM, Strecher VJ, Becker MH (1994) The health belief model and HIV risk behavior change. In: DiClemente RJ, Peterson JL (eds) Preventing AIDS: theories and methods of behavioral interventions. Plenum Press, New York

Scarinci IC, Bandura L, Hidalgo B, Cherrington A (2011) Development of a theory-based (PEN-3 and health belief model), culturally relevant intervention on cervical cancer prevention among Latina immigrants using intervention mapping. Health Promot Pract 12(3):

Strecher V, Rosenstock IM (1997) The health belief model. In: Glanz K, Lewis FM, Rimer BK, Viswanath K (eds) Health behavior and health education: theory, research and practice. Jossey-Bass, San Francisco

Urrutia MT (2009) Development and testing of a questionnaire: beliefs about cervical cancer and pap test in Chilean women. PhD Dissertation Thesis, University of Miami, Coral Gables, Miami

Weick KE (1979) The social psychology of organizing. McGraw-Hill, New York

Yin RK (2003) Case study research: design and methods, 3rd edn. Sage, Thousand Oats

Chapter 8
Overview on Multivariate Adaptive Regression Splines

Kweku-Muata Osei-Bryson

This chapter provides an overview of multivariate adaptive regression splines (MARS). Its main purpose is to introduce the reader to the major concepts underlying this data mining technique, particularly those that are relevant to the chapter that involves the use of this technique. This chapter includes an illustrative example and also provides guidance for interpreting a MARS model.

1 Introduction

Many research problems in IS and other areas involve the exploration of the relationship between candidate predictors (variables or factors) and the given target. In some cases, a predictive modeling technique (e.g., regression analysis) is used to address these problems. With such an approach, the researcher is required to hypothesize on the functional form of each candidate predictor as well as the interactions between the candidate predictors. Major decisions include

- Which predictor variables should be used?
- What is the mathematical form of $f(x_1, x_2, \ldots, x_M)$?
- What is the underlying functional form (e.g., log, square root, power, inverse, *S*-shape) for each predictor x_j?
- What interactions are to be considered and what is the appropriate degree of each interaction?
- How to explore conditional impacts?

Some approaches to data analysis, such as regression splines (RS) methods (e.g., Friedman 1991; Hastie and Tibshirani 1990), let the data dictate the functional

K.-M. Osei-Bryson (✉)
Department of Information Systems, Virginia Commonwealth University,
301 W. Main Street, Richmond, VA 23284, USA
e-mail: KMOsei@VCU.Edu

K.-M. Osei-Bryson and O. Ngwenyama (eds.), *Advances in Research Methods for Information Systems Research*, Integrated Series in Information Systems 34,
DOI: 10.1007/978-1-4614-9463-8_8,

form for each candidate predictor rather than requiring the researcher to specify it. Multivariate adaptive regression splines (MARS) induction automatically determines both variable selection and functional form, resulting in an explanatory predictive model. It offers representations of causal relationships using multiple functions over multiple regions of the decision space. It provides for the identification of conditional and nonlinear relationships. Thus, some questions that cannot be answered using regression analysis can be explored in greater depth using RS analysis. This feature of MARS analysis has resulted in its increased usage in IS research (e.g., Ko and Osei-Bryson 2004; Kositanurit et al. 2006; Mukkamala et al. 2006; Morawczynski and Ngwenyama 2007; Osei-Bryson et al. 2008; Ko et al. 2008; Hung et al. 2011). MARS has also been applied in various other fields including software engineering (e.g., Zhou and Leung 2007; Briand et al. 2004), natural language processing (Hu and Loizou 2008), computer-aided design (Behera et al. 2012), finance (e.g., Martin 2011; De Andrés et al. 2011), manufacturing (Guo et al. 2010), medicine (Deconinck et al. 2007), and biology (Balshi et al. 2009; Leathwick et al. 2005).

2 Regression Splines Modeling

MARS is a technique that may be used to discover, describe, and evaluate causal links between factors. Its development was influenced by the adaptive RS method of Hastie and Tibshirani (1990) and the recursive partition regression method of Breiman et al. (1984). While ordinary regression equations attempt to model the relationship between outcome and predictor variables using a single function, the RS approach models the relationship between the target (e.g., independent) and predictor variables as a linear combination of piecewise polynomial *basis functions* (BF) that are joined together smoothly at the knots.

2.1 Regression Splines Model

The RS approach models the relationship between outcome and predictor variables as a piecewise polynomial function $f(x)$ which can be obtained by dividing the range of each predictor variable into one or more intervals and representing f by a separate polynomial in each interval (Hastie et al. 2001). A regression spline function can be expressed as a linear combination of piecewise polynomial BF that are joined together smoothly at the knots (Fig. 1). When using the multivariate regression splines (MARS) for modeling the relationship between the predictor and dependent variables, it is not necessary to know the functional forms of the relationships, MARS establishes them based on the data. For example, in MARS, a general model of single predictor X_t and the dependent variable Y_t might take the form:

$$Y_t = \sum_{k=1}^{M} a_k B_k(X_t) + \varepsilon_t$$

Fig. 1 Illustration of knots in a regression splines plot

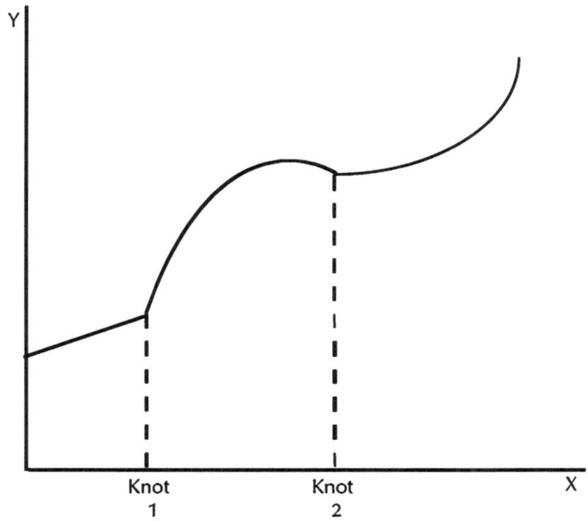

where $B_k (X_t)$ is the kth basis function of X_t. The coefficient of each BF (i.e., β_k) is estimated by minimizing the sum of square errors, which is similar to the coefficient estimation process of linear regression, but involving local data for the given region. MARS provides the analysis of variance (ANOVA) decomposition, which identifies the relative contributions of each of the predictor variables and the interactions between variables, and handles missing values.

2.2 Knots and Basis Function

For a given variable, a knot t marks the end of one region of data and the beginning of another:

- The behavior of the function changes at each knot
- MARS automatically generates knots based on the given data.

MARS uses both simple (elementary) BFs and complex BFs. Models with no interactions have elementary BFs only, while models that allow interactions between the predictor variables would include complex BFs.

Simple BFs involve a single variable (say x) and come in pairs of the form $(x - t)_+$ and $(t - x)_+$ where t is the knot, $(x - t)_+ = (x - t)$ if $x > t$, and 0 otherwise; and $(t - x)_+ = (t - x)$ if $x < t$, and 0 otherwise (Hastie and Tibshirani 1990; Hastie et al. 2001). In some cases, both forms occur in the final model such as BF1 and BF2 below:

$$\text{BF1} = \text{MAX(0, SHELL\_WEIGHT} - 0.155)$$
$$\text{BF2} = \text{MAX(0, 0.155} - \text{SHELL} - \text{WEIGHT)}$$

Complex BFs have the form: $h_k(\mathbf{x}) = \Pi_{\text{ij}} f_{ij} (x_i)$ where $x_1, \ldots x_q$ are the independent variables and f_{ij} is a BF for the ith independent variable x_i at jth knot. Each complex BF consists of the product of at least two elementary BFs. Models that allow interactions between variables include complex BFs such as BF9 below.

$$\begin{aligned} \text{BF9} &= \text{Max}(0, \text{SHUCKED\_WEIGHT} - 0.249) \times \text{BF2} \\ &= \text{Max}(0, \text{SHUCKED\_WEIGHT} - 0.249) \times \text{Max}(0, 0.155 - \text{SHELL\_WEIGHT}); \end{aligned}$$

Thus, BFs can represent either single-variable transformations or multivariable interaction terms.

2.3 Model Generation Process

The MARS approach builds a model in a two-phase process, using a forward-stepwise regression selection and backward-stepwise deletion strategy. In the forward phase, MARS builds an overfitted model by adding BFs. In the backward phase, BFs that have the least contribution to the model are deleted and the model is optimized. This process is repeated until all BFs have been eliminated. The end result of this deletion procedure is a unique sequence of candidate models. Similar to OLS regression, larger forward stage models will have higher R^2 values than smaller forward stage models but will also be more complex.

2.3.1 Forward Stage: Adding BFs

- MARS starts with just a constant in the initial model and then begins the search for a variable-knot combination (e.g., x_j, t_{jk}) that would result in the best improvement in the model where improvement is measured by the change in MSE. It should be noted that adding a BF always improves the MSE.
- This search process is then repeated in order to identify the best variable-knot combination to add, given the BFs already in the model.
- The brute search process continues until a user-specified limit on the number of BFs (i.e., Maximum Number of BFs) is reached.

2.3.2 Backward Stage: Deleting BFs

- Starting with the largest *forward stage* model, MARS determines the BF, which, using a residual sum of squares criteria, hurts the model the least if dropped;
- After refitting the now-pruned model, MARS again identifies a BF to drop (using a residual sum of squares criterion);
- This process is repeated until all BFs have been eliminated.
- The end result of this deletion procedure is a unique sequence of candidate models.

2.4 Selection of the Final Model

Selection of the final model may be based on the generalized cross-validation (GCV) criterion or on the closeness between the training MSE and the test MSE if a test subset of the data is used. The GCV can be expressed as follows:

$$\mathrm{GCV} = (1/N)\sum_{t=1}^{N}\left\{\left[Y_t - f_M(X_t)\right]^2/[1 - C(M)/N]^2\right\}$$

The numerator measures the lack of fit on the M basis function model $f_M\ (X_t)$, where there are N observations. This term corresponds to the sum of squared residuals from the fitted model. The denominator contains a penalty for model complexity, $C(M)$, which is related to the number of parameters estimated in the model. The GCV complexity measure is used to penalize larger models in a manner that is similar to the role of the Akaike Information Criterion (AIC), Schwarz–Bayes Criterion (SBC), or Bayes Information Criterion (BIC) for regression models. This penalty is based on the degrees of freedom (df) charged per knot, which can be specified manually or determined automatically using methods such as random selection of a portion of the data for testing, or tenfold cross-validation.

3 Illustrative Example

Our illustrative example involves using the Abalone dataset to develop a model that would predict the value of the RINGS variable. We display a set of captioned figures and tables that describes major steps and results of the process (Figs. 2, 3, 4, 5, 6, 7, 8, 9, 10 and Table 1).

4 Interpreting a MARS Model

There is often an interest in understanding the impacts of a given predictor variable on the target variable. Below we refer to this predictor variable as the subject predictor variable. These impacts can be determined using the following steps:

1. Identify and select the BFs in the RS equation that are associated with the given subject predictor variable. For example, if SHELL_WEIGHT was the subject predictor variable, then the relevant BFs would be BF2, BF9, B16, BF17, BF18 (see Table 2).
2. Given the selected BFs, identify the knots of the given subject predictor variable (e.g., 0.155 for SHELL_WEIGHT).
3. Given the set of knots, identify the set of intervals. If there is a single knot (e.g., for SHELL_WEIGHT), then there are 2 intervals ($\leq$0.155, >0.155). If there are

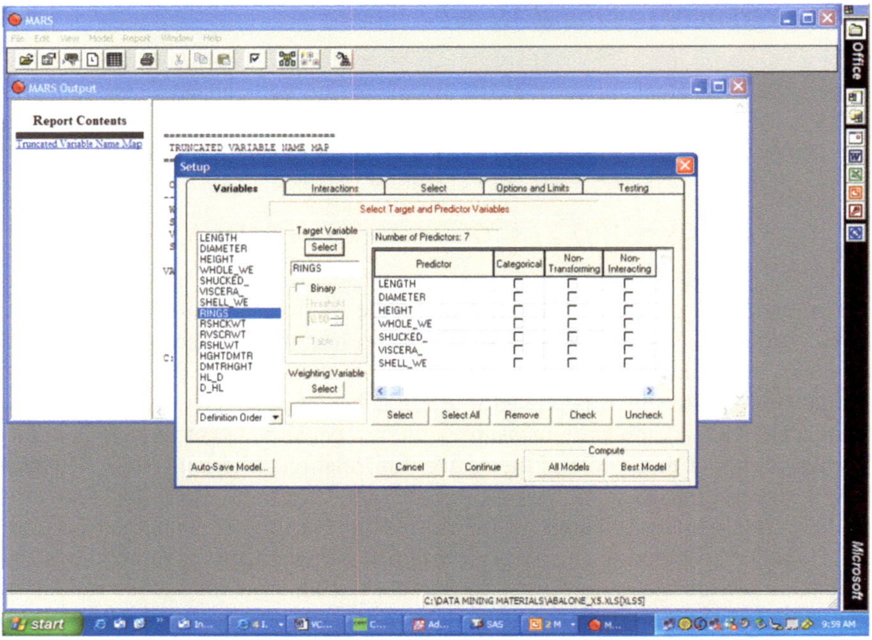

Fig. 2 Specifying target and input variables

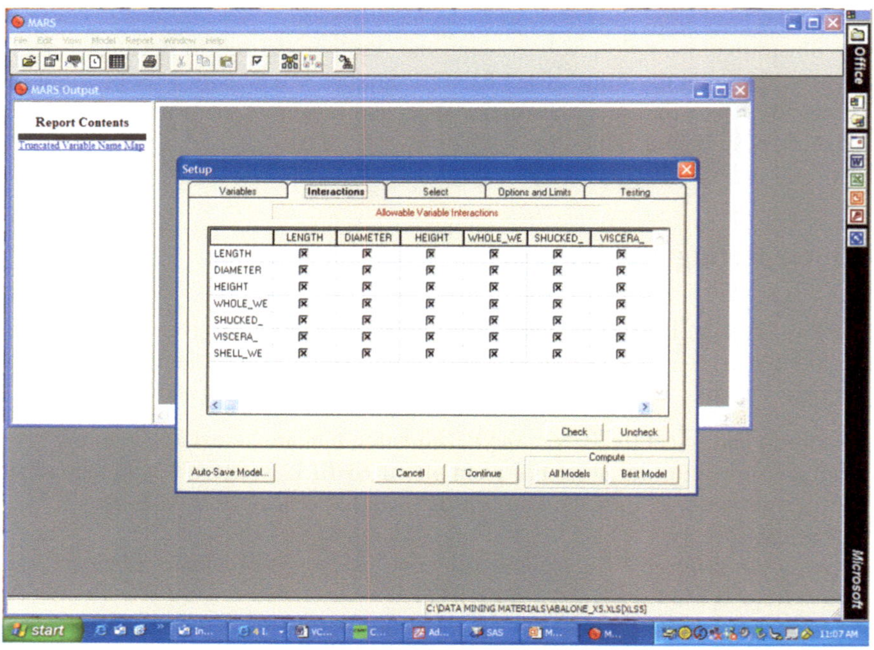

Fig. 3 Specifying two-way interactions between variables

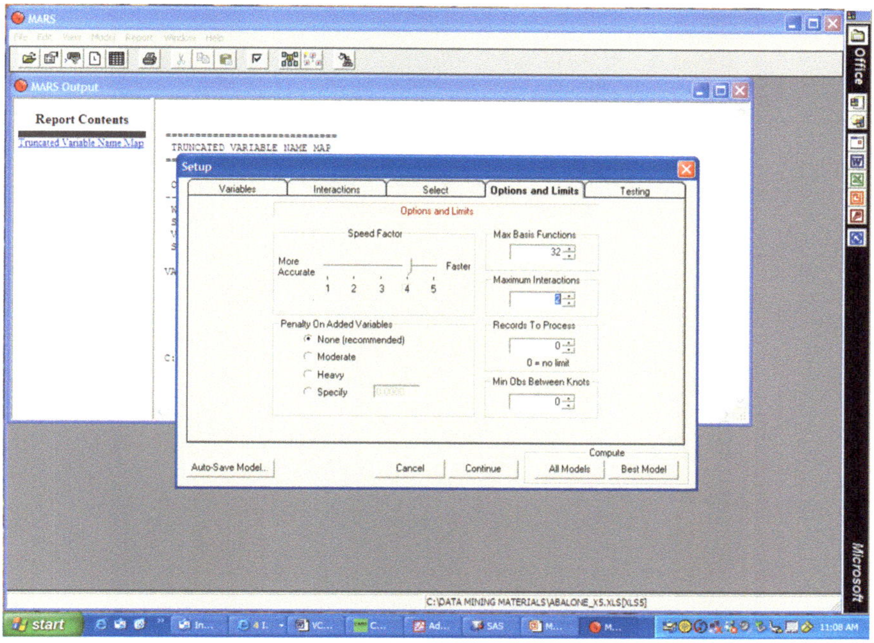

Fig. 4 Specifying parameters

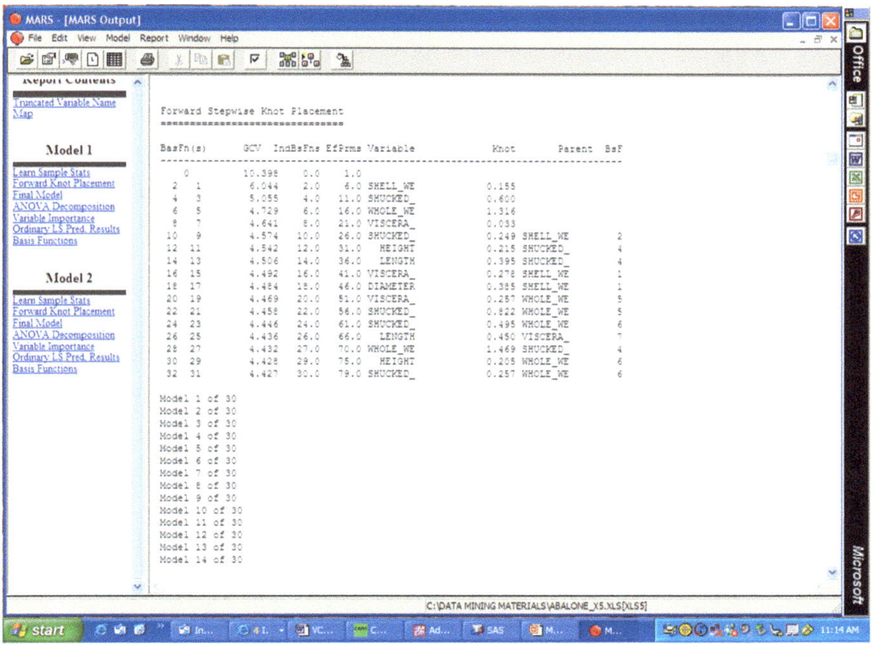

Fig. 5 Forward phase knot placement

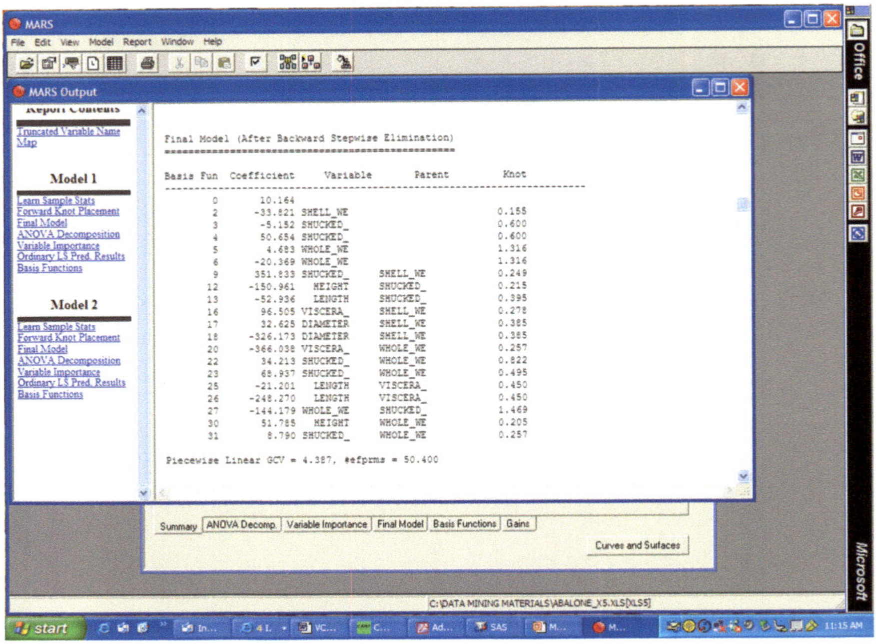

Fig. 6 Final model—backward phase knot placement

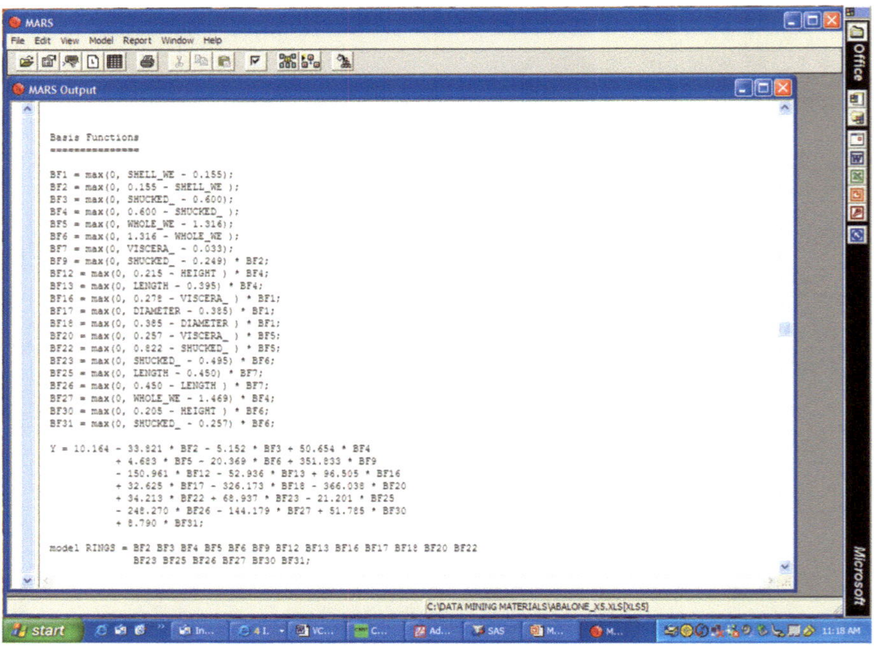

Fig. 7 Final model—basis functions (BFs)

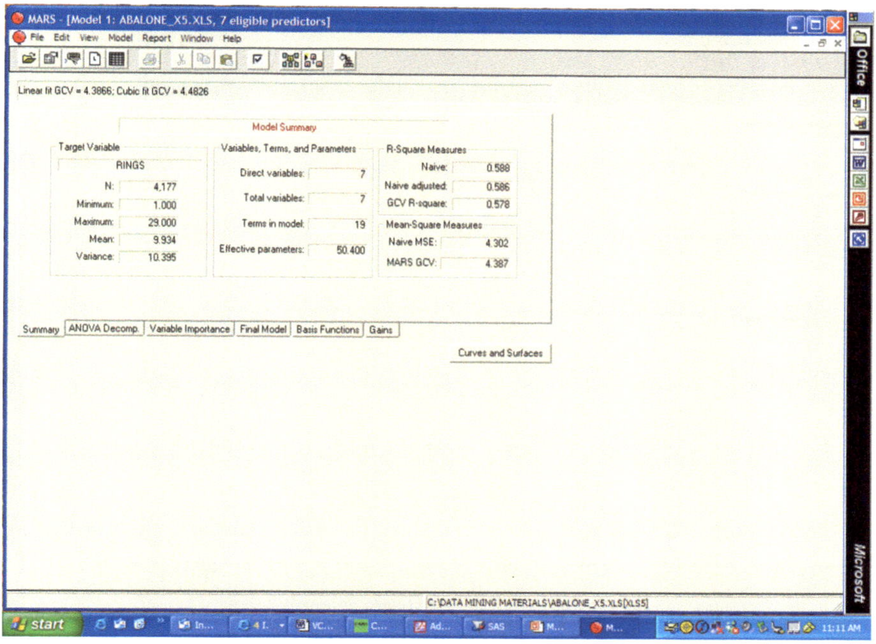

Fig. 8 Final model—summary statistics

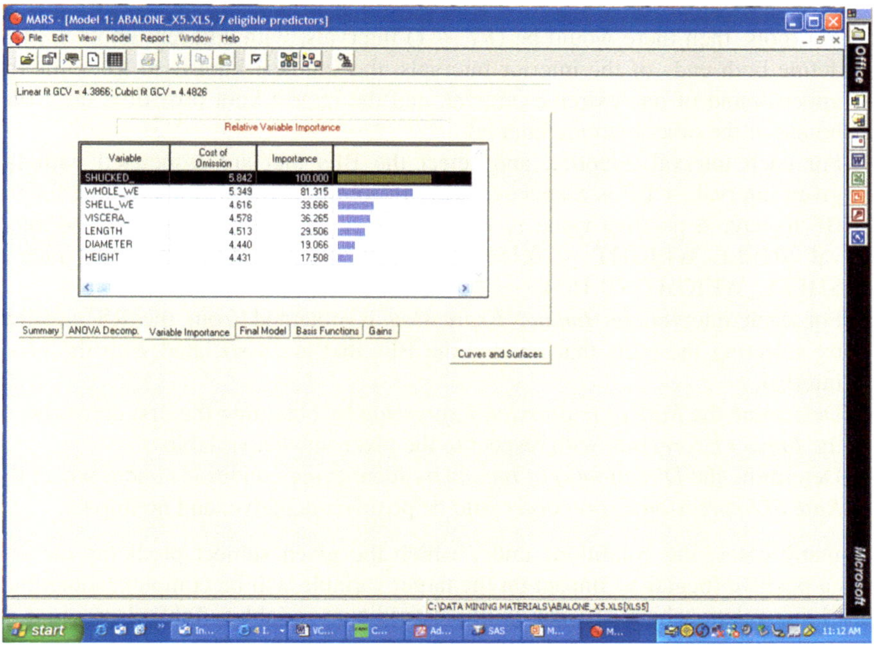

Fig. 9 Final model—variable importance

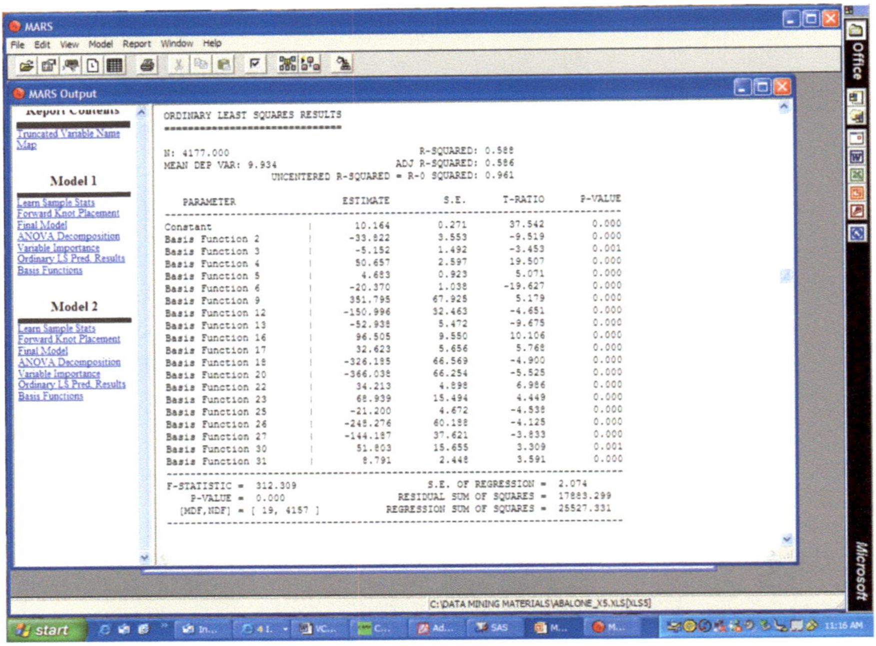

ORDINARY LEAST SQUARES RESULTS

N: 4177.000 R-SQUARED: 0.588
MEAN DEP VAR: 9.934 ADJ R-SQUARED: 0.586
UNCENTERED R-SQUARED = R-0 SQUARED: 0.961

PARAMETER	ESTIMATE	S.E.	T-RATIO	P-VALUE
Constant	10.164	0.271	37.542	0.000
Basis Function 2	-33.822	3.553	-9.519	0.000
Basis Function 3	-5.152	1.492	-3.453	0.001
Basis Function 4	50.657	2.597	19.507	0.000
Basis Function 5	4.683	0.923	5.071	0.000
Basis Function 6	-20.370	1.038	-19.627	0.000
Basis Function 9	351.795	67.925	5.179	0.000
Basis Function 12	-150.996	32.463	-4.651	0.000
Basis Function 13	-52.938	5.472	-9.675	0.000
Basis Function 16	96.505	9.550	10.106	0.000
Basis Function 17	32.623	5.656	5.768	0.000
Basis Function 18	-326.185	66.569	-4.900	0.000
Basis Function 20	-366.038	66.254	-5.525	0.000
Basis Function 22	34.213	4.898	6.986	0.000
Basis Function 23	68.939	15.494	4.449	0.000
Basis Function 25	-21.200	4.672	-4.538	0.000
Basis Function 26	-248.276	60.188	-4.125	0.000
Basis Function 27	-144.187	37.621	-3.833	0.000
Basis Function 30	51.803	15.655	3.309	0.001
Basis Function 31	8.791	2.448	3.591	0.000

F-STATISTIC = 312.309 S.E. OF REGRESSION = 2.074
P-VALUE = 0.000 RESIDUAL SUM OF SQUARES = 17883.299
[MDF,NDF] = [19, 4157] REGRESSION SUM OF SQUARES = 25527.331

Fig. 10 Final model—ordinary least squares statistics

K_V knots, then there would be $(K_V + 1)$ intervals, with adjacent knots used to define both ends of the interior intervals, the smallest knot used to define the upper bound of one extreme interval, and the largest knot providing the lower bound of the other extreme interval.

4. For each interval, identify, and select the BFs that are associated with the given interval. A BF is associated with a given interval, if it is possible for the BF to have a positive value in the interval (e.g., BF2 and BF9 for the interval SHELL_WEIGHT $\leq$ 0.155; BF16, BF17, and BF18 for the interval SHELL_WEIGHT > 0.155).
5. For each interval, its *Impact Expression* is obtained from the RS equation by selecting the parts that involve the BFs that are associated with the given interval.
6. Determine the *Rate of Impression Expression* by obtaining the first derivative of the *Impact Expression* with respect to the given subject variable.
7. Determine the *Direction(s) of Impact* by identifying conditions under which the *Rate of Impression Expression* could be positive, negative, and no impact.

In many cases, the conditions under which the given subject predictor variable has a positive (negative) impact on the target variable will be complex, sometimes involving relationships between multiple predictor variables. Table 3 provides an example that involves the predictor variable SHELL_WEIGHT as the subject predictor variable.

Table 1 Final model—expanded description of complex basis functions

BF	Raw expression	Expanded expression
BF1	Max (0, SHELL_WEIGHT – 0.155)	
BF2	Max (0, 0.155 – SHELL_WEIGHT)	
BF3	Max (0, SHUCKED_WEIGHT – 0.600)	
BF4	Max (0, 0.600 – SHUCKED_WEIGHT)	
BF5	Max (0, WHOLE_WEIGHT – 1.316)	
BF6	Max (0, 1.316 – WHOLE_WEIGHT)	
BF7	Max (0, VISCERA_WEIGHT – 0.033)	
BF9	Max (0, SHUCKED_WEIGHT – 0.249) × BF2	Max (0, SHUCKED_WEIGHT – 0.249) × Max (0, 0.155 – SHELL_WEIGHT)
BF12	Max (0, 0.215 – HEIGHT) × BF4	Max (0, 0.215 – HEIGHT) × Max (0, 0.600 – SHUCKED_WEIGHT)
BF13	Max (0, LENGTH – 0.395) × BF4	Max (0, LENGTH – 0.395) × Max (0, 0.600 – SHUCKED_WEIGHT)
BF16	Max (0, 0.270 – VISCERA_WEIGHT) × BF1	Max (0, 0.270 – VISCERA_WEIGHT) × Max (0, SHELL_WEIGHT – 0.155)
BF17	Max (0, DIAMETER – 0.385) × BF1	Max (0, DIAMETER – 0.385) × Max (0, SHELL_WEIGHT – 0.155)
BF18	Max (0, 0.385 – DIAMETER) × BF1	Max (0, 0.385 – DIAMETER) × Max (0, SHELL_WEIGHT – 0.155)
BF20	Max (0, 0.257 – VISCERA_WEIGHT) × BF5	Max (0, 0.257 – VISCERA_WEIGHT) × Max (0, WHOLE_WEIGHT – 1.316)
BF22	Max (0, 0.822 – SHUCKED_WEIGHT) × BF5	Max (0, 0.822 – SHUCKED_WEIGHT) × Max (0, WHOLE_WEIGHT – 1.316)
BF23	Max (0, SHUCKED_WEIGHT – 0.495) × BF6	Max (0, SHUCKED_WEIGHT – 0.495) × Max (0, 1.316 – WHOLE_WEIGHT)
BF25	Max (0, LENGTH – 0.450) × BF7	Max (0, LENGTH – 0.450) × Max (0, VISCERA_WEIGHT – 0.033)
BF26	Max (0, 0.450 – LENGTH) × BF7	Max (0, 0.450 – LENGTH) × Max (0, VISCERA_WEIGHT – 0.033)
BF27	Max (0, WHOLE_WEIGHT – 1.469) × BF4	Max (0, WHOLE_WEIGHT – 1.469) × Max (0, 0.600 – SHUCKED_WEIGHT)
BF30	Max (0, 0.205 – HEIGHT) × BF6	Max (0, 0.205 – HEIGHT) × Max (0, 1.316 – WHOLE_WEIGHT)
BF31	Max (0, SHUCKED_WEIGHT – 0.257) × BF6	Max (0, SHUCKED_WEIGHT – 0.257) × Max (0, 1.316 – WHOLE_WEIGHT)

Table 2 Final model—regression splines equation described using predictor variables

RINGS		
=	10.164	
	−33.611 × BF2	−33.611 × Max (0, 0.155 − SHELL_WEIGHT)
	−51.152 × BF3	−51.152 × Max (0, SHUCKED_WEIGHT − 0.600)
	+50.654 × BF4	+50.654 × Max (0, 0.600 − SHUCKED_WEIGHT)
	−4.683 × BF5	−4.683 × Max (0, WHOLE_WEIGHT − 1.316)
	−20.369 × BF6	−20.369 × Max (0, 1.316 − WHOLE_WEIGHT)
	+357.883 × BF9	+357.883 × Max (0, SHUCKED_WEIGHT − 0.249) × Max (0, 0.155 − SHELL_WEIGHT)
	−150.961 × BF12	−150.961 × Max (0, 0.215 − HEIGHT) × Max (0, 0.600 − SHUCKED_WEIGHT)
	−51.936 × BF13	−51.936 × Max (0, LENGTH − 0.395) × Max (0, 0.600 − SHUCKED_WEIGHT)
	+96.505 × BF16	+96.505 × Max (0, 0.270 − VISCERA_WEIGHT) × Max (0, SHELL_WEIGHT − 0.155)
	+32.625 × BF17	+32.625 × Max (0, DIAMETER − 0.385) × Max (0, SHELL_WEIGHT − 0.155)
	−326.173 × BF18	−326.173 × Max (0, 0.385 − DIAMETER) × Max (0, SHELL_WEIGHT − 0.155)
	−366.038 × BF20	−366.038 × Max (0, 0.257 − VISCERA_WEIGHT) × Max (0, WHOLE_WEIGHT − 1.316)
	+34.213 × BF22	+34.213 × Max (0, 0.822 − SHUCKED_WEIGHT) × Max (0, WHOLE_WEIGHT − 1.316)
	+68.937 × BF23	+68.937 × Max (0, SHUCKED_WEIGHT − 0.495) × Max (0, 1.316 − WHOLE_WEIGHT)
	−21.201 × BF25	−21.201 × Max (0, LENGTH − 0.450) × Max (0, VISCERA_WEIGHT − 0.033)
	−248.270 × BF26	−248.270 × Max (0, 0.450 − LENGTH) × Max (0, VISCERA_WEIGHT − 0.033)
	−144.179 × BF27	−144.179 × Max (0, WHOLE_WEIGHT − 1.469) × Max (0, 0.600 − SHUCKED_WEIGHT)
	+51.785 × BF30	+51.785 × Max (0, 0.205 − HEIGHT) × Max (0, 1.316 − WHOLE_WEIGHT)
	+8.790 × BF31	+8.790 × Max (0, SHUCKED_WEIGHT − 0.257) × Max (0, 1.316 − WHOLE_WEIGHT)

Table 3 Impact of SHELL_WEIGHT on RINGS

Int	Impact expression	Rate of impact expression	Direction(s) of impact
≤0.155	−33.611 × (0.155 − SHELL_WEIGHT) + 357.883 × Max (0, SHUCKED_WEIGHT − 0.249) × (0.155 − SHELL_WEIGHT)	33.611 − 357.883 × Max (0, SHUCKED_WEIGHT − 0.249)	*Positive* If SHUCKED_WEIGHT < 0.249 + (33.611 /357.883); *Negative* Otherwise
>0.155	+96.505 × Max (0, 0.270 − VISCERA_ WEIGHT) × (SHELL_WEIGHT − 0.155) +32.625 × Max (0, DIAMETER − 0.385) × (SHELL_ WEIGHT − 0.155) −326.173 × Max (0, 0.385 − DIAMETER) × (SHELL_ WEIGHT − 0.155)	+96.505 × Max (0, 0.270 − VISCERA_WEIGHT) +32.625 × Max (0, Diameter − 0.385) −326.173 × Max (0, 0.385 − DIAMETER)	*Positive* If Diameter ≥ 0.385; *Negative* If (Diameter < 0.385) and (VISCERA_WEIGHT ≥ 0.270);

5 Conclusion

It should be noted that both regression and RS can identify the order of importance of the independent variables in a predictive model and estimate the value of the coefficient for each independent variable. However, if the impact of an independent variable on the dependent variable is conditional, then RS can identify such conditions, while regression cannot. Thus, some questions cannot be answered using regression since it does not provide means for exploring those questions. On the other hand, RS can provide the means for exploring some research questions in greater depth than would have been possible using regression.

References

Balshi MS, McGuire AD, Duffy P, Flannigan M, Walsh J, Melillo J (2009) Assessing the response of area burned to changing climate in Western Boreal North America using a multivariate adaptive regression splines (MARS) approach. Glob Change Biol 15(3):578–600

Behera AK, Verbert J, Lauwers B, Duflou JR (2012) Tool path compensation strategies for single point incremental sheet forming using multivariate adaptive regression splines. Comput-Aided Des 45(3):575–590

Breiman L, Friedman J, Olshen R, Charles S (1984) Classification and regression trees. Wadsworth International Group

Briand L, Freimut B, Vollei F (2004) Using multiple adaptive regression splines to understand trends in inspection data and identify optimal inspection rates. J Syst Softw 73(2):2–23

De Andrés J, Lorca P, de Cos Juez FJ, Sánchez-Lasheras F (2011) Bankruptcy forecasting: a hybrid approach using fuzzy c-means clustering and multivariate adaptive regression splines (MARS). Expert Syst Appl 38(3):1866–1875

Deconinck E, Coomans D, Vander Heyden Y (2007) Exploration of linear modelling techniques and their combination with multivariate adaptive regression splines to predict gastro-intestinal absorption of drugs. J Pharm Biomed Anal 43(1):119–130

Friedman JH (1991) Multivariate Adaptive Regression Splines. Ann Stat 19(1):1, pp 1–141

Guo W, Zhao N, Shao H (2010) IT investment efficiency analysis of equipment manufacturing industry based on two-stage nonparametric model. In: Proceedings of IEEE 2010 international conference on challenges in environmental science and computer engineering, vol 2, pp 21–24

Hastie T, Tibshirani R (1990) Generalized additive model. Chapman and Hall, London

Hastie T, Tibshirani R, Friedman J (2001) The elements of statistical learning: data mining, inference, and prediction. Springer-Verlag, New York

Hu Y, Loizou PC (2008) Evaluation of objective quality measures for speech enhancement. IEEE Trans Audio Speech Lang Process 16(1):229–238

Hung Y-H, Chou S-C, Tzeng G-H (2011) Knowledge management adoption and assessment for SMES by a novel MCDM approach. Decis Support Syst 51:270–291

Ko M, Osei-Bryson K (2004) Using regression splines to assess the impact of information technology investments on productivity in the healthcare industry. Inf Syst J 14:43–63

Ko M, Clark JG, Ko D (2008) Revisiting the impact of information technology investments on productivity: an empirical investigation using multivariate adaptive regression splines. Inf Res Manage J 21(3):1–23

Kositanurit B, Ngwenyama O, Osei-Bryson K-M (2006) An exploration of factors that impact individual performance in an ERP environment: an analysis using multiple analytical techniques. EurJ Inf Syst 15:556–568

Leathwick JR, Rowe D, Richardson J, Elith J, Hastie T (2005) Using multivariate adaptive regression splines to predict the distributions of New Zealand's freshwater diadromous fish. Freshw Biol 50(12):2034–2052

Morawczynski O, Ngwenyama O (2007) Unraveling the impact of investments in ICT, education and health on development: an analysis of archival data of five West African countries using regression splines. Electron J Inf Syst Dev Countries 29:1–15

Mukkamala S, Sung AH, Abraham A, Ramos V (2006) Intrusion detection systems using adaptive regression spines. In: Enterprise Information Systems VI, pp 211–218. Springer, Netherlands

Osei-Bryson K-M, Dong L, Ngwenyama O (2008) Exploring managerial factors affecting ERP implementation: an investigation of the Klein-Sorra model using regression splines. Inf Syst J 18(5):499–527

Martin A (2011) A Hybrid model for bankruptcy prediction using genetic algorithm, fuzzy c-means and MARS. Int J Soft Comput 2(1):12–24

Zhou Y, Leung H (2007) Predicting object-oriented software maintainability using multivariate adaptive regression splines. J Syst Softw 80(8):1349–1361

Chapter 9
Reexamining the Impact of Information Technology Investments on Productivity Using Regression Tree and MARS-Based Analyses

Myung Ko and Kweku-Muata Osei-Bryson

Several studies have investigated the impact of investments in IT on productivity. In this chapter, we revisit this issue and reexamine the impact of investments in IT on hospital productivity using two data mining techniques, which allowed us to explore interactions between the input variables as well as conditional impacts. The results of our study indicated that the relationship between IT investment and productivity is very complex. We found that the impact of IT investment is not uniform and the rate of IT impact varies contingent on the amounts invested in the *IT Stock*, *Non-IT Labor*, *Non-IT Capital*, and possibly time.

1 Introduction

Evaluating the true impact of IT on organizations has been a constant concern for both researchers and practitioners and numerous studies have investigated this issue for more than 3 decades. Several recent studies have reported that a positive relationship between IT and productivity at the firm level (e.g., Lichtenberg 1995; Brynjolfsson and Hitt 1996; Hitt and Brynjolfsson 1996; Dewan and Min 1997; Mukopadhyay et al. 1997; Menon et al. 2000; Shao and Lin 2001; Kudyba and Diwan 2002; Shin 2006). While previous studies made significant contribution to IT & productivity research, these studies focused on examining the impact of

M. Ko (✉)
Department of Information Systems and Cyber Security One UTSA Circle, The University of Texas at San Antonio, San Antonio, TX 78249, USA
e-mail: Myung.Ko@utsa.edu

K.-M. Osei-Bryson
Department of Information Systems, Virginia Commonwealth University, 301 W. Main Street, Richmond, VA 23284, USA
e-mail: KMOsei@VCU.Edu

K.-M. Osei-Bryson and O. Ngwenyama (eds.), *Advances in Research Methods for Information Systems Research*, Integrated Series in Information Systems 34,
DOI: 10.1007/978-1-4614-9463-8_9,

Table 1 Capabilities of data mining techniques

Capability	RT	MARS
Ability to detect interactions	Yes	Yes
Estimate values of the coefficient for each variable	No	Yes
Identify order of importance of variable	Yes	Yes
Ability to build a model by partitioning a variable	Yes	Yes

IT investment on productivity in terms of its existence or nonexistence. We suggest that, at current stage of IT & productivity research, the appropriate research question that should be addressed is not "does IT impact productivity?" but "under what conditions do investments in IT impact productivity?"

In this study, we use two popular data mining techniques, regression trees (RT) and multivariate adaptive regression splines (MARS), to explore our research question. Our reasons for using this pair of techniques are primarily because they have capabilities for discovering nonlinear relationship between the response and predictor variables (Deichmann et al. 2002) and identifying interactions and conditional relationships between the predictor variables. Further, neither of these approaches require the use of a parametric theoretical model. Table 1 describes the capabilities of each technique used in our analysis.

2 Empirical Analysis

2.1 Description of the Data

For this study, we use a dataset that has been used in previous studies on the impact of investments in the IT Stock on organizational productivity (Menon et al. 2000; Menon and Lee 2000). The description of data used in this study is included in Table 2. The data were collected by the Department of Health. They include all hospitals in the state of Washington except for specialized hospitals for the period and cover from 1976 to 1994. The data included 1,130 data points, and each data point represents a hospital. It is unbalanced panel data. Thus, each hospital may or may not be presented all years. The input variables are Diagnosis-Related Grouping (*DRG*), *IT Stock* (*T*), *Non-IT Labor* (*L*), and *Non-IT Capital* (*K*).

In this study, we use the log value for each continuous input variable because several theoretical production functions [i.e., Cobb-Douglas, constant elasticity of substitution (CES), and translog] that are commonly used in IT and productivity research (e.g., Menon et al. 2000; Hitt and Brynjolfsson 1996) involve the log of the variable rather than the raw variable. Table 3 describes the summary statistics of the data used in this study.

Table 2 Variable definitions (*Source* Menon et al. (2000) and Menon's SAS Program)

Variable	Description (or departmental account)
Adjusted patient days (*V*)	Sum of *Inpatient Days* and *Outpatient Days*. Deflated by the output price (see below)
DRG	This is a binary variable that indicates whether the Diagnosis Related Group (DRG) applies for the given observation. The DRG was implemented in 1983 and so DRG = 1 if the observation year is 1983 or later, and DRG = 0 otherwise
IT Stock (*T*)	Represents IT investments and is calculated as *IT Capital* plus *Medical IT Capital* plus three times *IT Labor*. This formula is consistent with the approach of Hitt and Brynjolfsson (Groff et al. 2001) that *IT Labor* is considered to be a capital asset
IT Capital	IT Capital expenses incurred mainly for administrative purposes in the departmental accounts. Deflated by Price Deflator for Fixed Investment for IT from WEFA (1994)
IT Labor	Salaries and employee benefits incurred in the IT Capital accounts. Deflated by Labor Price (see below)
Medical IT Capital	Capital expenses incurred for the equipment used for diagnosing and therapeutics in the departmental accounts. Deflated by Price Deflator for Fixed Investment for IT from WEFA (1994)
Non-IT Capital (*K*)	Capital expenses incurred for the equipment used for therapeutics purposes only and any capital expenses in remaining departmental accounts. Deflated by Price Deflator for Fixed Investment for Non-IT from WEFA (1994)
Non-IT Labor (*L*)	Salaries, employee benefits, and physicians' salaries charged to accounts other than IT Capital accounts. Deflated by Labor Price (see below)
Labor price	Employment Price Index for healthcare services from Bureau of Labor Statistics (BLS) (1995)
Output price	Consumer Price Index for healthcare services from WEFA (1994)

Table 3 Summary statistics (log value)

Variables	Min	Max	Mean	Std. dev.
Adjusted patient days (*V*)	8.292	12.561	10.514	0.850
Non-IT capital (*K*)	10.010	16.499	13.266	1.266
Non-IT labor (*L*)	13.473	18.659	16.136	0.965
IT stock (*T*)	13.042	18.195	15.526	0.904

2.2 Results of Regression Tree-Based Analysis

We generated a RT using the CART 5.0 software based on a tenfold cross-validation method. We set Minimum Number of Observation per Leaf to 30 and the Minimum Number of Observations for a Split Search to 60 in order. The resulting RT model had an R-squared of 0.829, suggesting that it had high predictive power. Figure 1 provides an overview graphic description of the topology of the resulting RT and allows us to identify predictor variables. Table 5 (a and b) describe the rule set, with the difference in the two just being how the rules are sorted (Table 4).

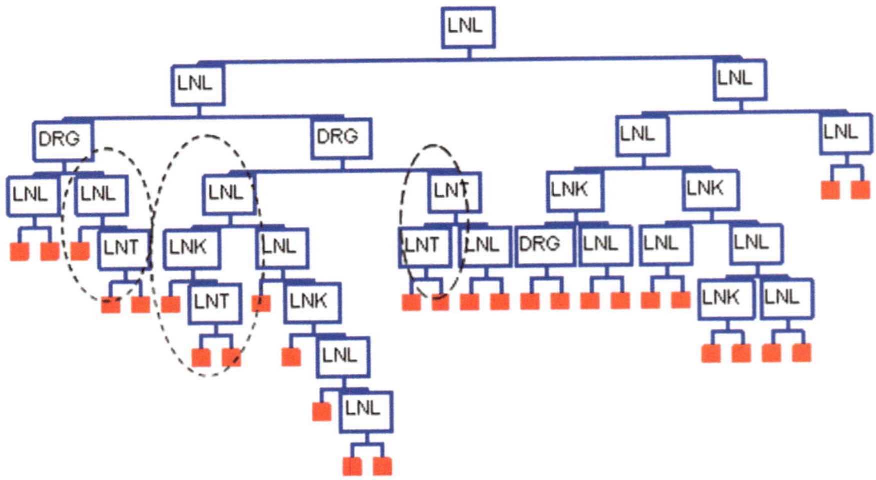

Fig. 1 Regression tree (RT) topology—predictor variables

Table 4 Relative importance of variables by RT

Variable	Importance
Non-IT Labor ($\log_e L$)	100.000
IT Stock ($\log_e T$)	86.48
Non-IT Capital ($\log_e K$)	48.97
DRG	11.39

With regard to the impact of investments in the *IT Stock* on productivity, this set of rules suggests that.

1. Investments in the *IT Stock* have an impact on productivity only when CART rules 4, 5, 7, 8, 14, 15, 16, and 17 apply, since the corresponding rules include the *IT Stock* (*T*) as a predictor. However, the investments in the *IT Stock* do not appear to have an impact on productivity when the rest of the rules apply since the corresponding rules do not include *IT Stock* (*T*) as a predictor.
2. When investments in the *IT Stock* have an impact on productivity, the rate of this impact appears to be positive (e.g., rules 4 & 5, 7 & 8, and 14–17).
3. The impact of investments in IT on productivity is contingent on the amounts invested in *Non-IT Labor* (*L*) and/or *Non-IT Capital* (*K*) (see Fig. 1, and Table 5 (a, b).
4. The impact of investments in the *IT Stock* (*T*) differs when the *DRG* applies (i.e., *DRG* = 1) than when it did not apply (i.e., *DRG* = 0). For example, when the *DRG* applies, whether or not investments in the *IT Stock* (*T*) have an impact on productivity is contingent on both the amounts invested in *Non-IT Labor* (*L*) and *Non-IT Capital* (*K*) (see rules 7, 8). On the other hand, when the *DRG* does not apply, it is possible that investments in the *IT Stock* (*T*) can have an impact on

Table 5 (**a**) RT rules, sorted by DRG, terminal node ID. (**b**) RT Rules, sorted by $\log_e V$

Rule ID	Input variables				Output:
	DRG	$\log_e L$	$\log_e K$	$\log_e T$	Mean($\log_e V$)
(**a**)					
3	0	≤14.4707			9.0758
4	0	(14.4707, 15.1748]		≤14.4348	9.7691
5	0	(14.4707, 15.1748]		>14.4348	10.0208
14	0	(15.1748, 16.3942]		≤14.881	10.2444
15	0	(15.1748, 16.3942]		(14.881, 15.0575]	10.5331
16	0	(15.1748, 15.9294]		>15.0575	10.7343
17	0	(15.9294, 16.3942]		>15.0575	11.0474
19	0	(16.3942, 16.8873]	≤13.6985		11.3903
1	1	≤14.779			8.7406
2	1	(14.779, 15.1748]			9.2107
6	1	(15.1748, 15.6695]	≤12.8136		9.8354
7	1	(15.1748, 15.6695]	>12.8136	≤15.0282	9.4630
8	1	(15.1748, 15.6695]	>12.8136	>15.0282	9.6945
9	1	(15.6695, 15.8568]			10.0494
10	1	(15.8568, 16.3942]	≤13.191		10.4776
11	1	(15.8568, 16.0988]	>13.191		10.1665
12	1	(16.0988, 16.2352]	>13.191		10.2866
13	1	(16.2352, 16.3942]	>13.191		10.3608
18	1	(16.3942, 16.8873]	≤13.6985		11.1201
20	1_or_0	(16.3942, 16.6099]	>13.6985		10.6539
21	1_or_0	(16.6099, 16.8873]	>13.6985		10.9445
22	1_or_0	(16.8873, 17.0842]	≤14.3351		11.4283
23	1_or_0	(17.0842, 17.6502]	≤14.3351		11.6842
24	1_or_0	(16.8873, 17.2691]	(14.3351, 14.485]		11.1287
25	1_or_0	(16.8873, 17.2691]	>14.485		10.9927
26	1_or_0	(17.2691, 17.5001]	>14.351		11.3332
27	1_or_0	(17.5001, 17.6502]	>14.351		11.4299
28	1_or_0	(17.6502, 17.859]			11.7402
29	1_or_0	>17.859			12.048
(**b**)					
1	1	≤14.779			8.7406
3	0	≤14.4707			9.0758
2	1	(14.779, 15. 1748]			9.2107
7	1	(15.1748, 15. 6695]	>12.8136	≤15.0282	9.4630
8	1	(15.1748, 15. 6695]	>12.8136	>15.0282	9.6945
4	0	(14.4707, 15. 1748]		≤14.4348	9.7691
6	1	(15.1748, 15. 6695]	≤12.8136		9.8354
5	0	(14.4707, 15. 1748]		>14.4348	10.0208
9	1	(15.6695, 15.8568]			10.0494
11	1	(15.8568, 16.0988]	>13.191		10.1665
14	0	(15.1748, 16.3942]		≤14.881	10.2444
12	1	(16.0988, 16.2352]	>13.191		10.2866

(continued)

Table 5 (continued)

Rule ID	Input variables				Output: Mean($\log_e V$)
	DRG	$\log_e L$	$\log_e K$	$\log_e T$	
13	1	(16.2352, 16.3942]	>13.191		10.3608
10	1	(15.8568, 16.3942]	≤13.191		10.4776
15	0	(15.1748, 16.3942]		(14.881, 15.0575]	10.5331
20	1_or_0	(16.3942, 16.6099]	>13.6985		10.6539
16	0	(15.1748, 15.9294]		>15.0575	10.7343
21	1_or_0	(16.6099, 16.8873]	>13.6985		10.9445
25	1_or_0	(16.8873, 17.2691]	>14.485		10.9927
17	0	(15.9294, 16.3942]		>15.0575	11.0474
18	1	(16.3942, 16.8873]	≤13.6985		11.1201
24	1_or_0	(16.8873, 17.2691]	(14.3351, 14.485]		11.1287
26	1_or_0	(17.2691, 17.5001]	>14.351		11.3332
19	0	(16.3942, 16.8873]	≤13.6985		11.3903
22	1_or_0	(16.8873, 17.0842]	≤14.3351		11.4283
27	1_or_0	(17.5001, 17.6502]	>14.351		11.4299
23	1_or_0	(17.0842, 17.6502]	≤14.3351		11.6842
28	1_or_0	(17.6502, 17.859]			11.7402
29	1_or_0	>17.859			12.048

"(": greater than or equal to; "]": less than

productivity is contingent on the amount invested in *Non-IT Labor* (*L*) but independent of the amount invested in *Non-IT Capital* (*K*) (see rules 4–5, 14–17).

5. The results in Table 5b suggest that the highest levels of productivity can be achieved without the impact of IT investment since the rules associated with the highest mean values for the target variable (e.g., rules 23, 28, and 29) do not include investments in the *IT Stock* (*T*) as a predictor. Thus, it would appear that the IT productivity paradox could occur as additional amounts invested in the *IT Stock* (*T*) would have no impact on productivity.

Overall, these results suggest that *IT Stock* has an impact on productivity, but the rate of this impact is in fact contingent on the amounts invested in *Non-IT Labor*, the *IT Stock*, and/or *Non-IT Capital*, as well as to whether the *DRG* applies. Thus, the results also suggest that it is possible for the productivity paradox to occur. Further, with regard to the issue of whether investments in IT should be considered as a substitute to investments in *Non-IT Labor*, our RT results suggest that in general, such a position does not hold.

2.3 Results of MARS Analysis

We used the MARS 2.0 software to generate our regression splines model using the following parameter settings: Maximum Number Variables in an

Table 6 Relative importance of variables by MARS

Variable	Importance
Non-IT Labor ($\log_e L$)	100.000
DRG	65.916
Non-IT Capital ($\log_e K$)	31.695
IT Stock ($\log_e T$)	28.419

Interaction = 3; Minimum Number of Observations between Knots = 10; Maximum Number of Basis Functions = 24 BFs; Testing: Every fifth observation is used. The resulting regression splines model had an R-squared of 0.945, indicating that it is a fairly strong model. Table 6 displays the order of relative importance of the input variables, indicating that investments in the *IT Stock* (*T*) are the least important of the four potential predictors.

Table 7 describes the basis functions of the MARS model. The first term included is always a constant (basis function 0). As shown in Table 7, some basis functions (e.g., BF4 and BF13) do not have their own coefficient since they only exist as part of other basis functions (e.g., BF9, and BF15). Based on the model shown in Table 7, our MARS model can be expressed as follows:

$$\begin{aligned}\log_e V = {} & 10.522 + 0.641 \times \text{BF1} - 0.886 \times \text{BF2} + 0.982 \times \text{BF3} + 0.366 \times \text{BF5} \\ & - 0.788 \times \text{BF9} + 0.899 \times \text{BF10} - 2.388 \times \text{BF15} - 0.160 \times \text{BF16} \\ & + 0.335 \times \text{BF17} + 0.171 \times \text{BF18}\end{aligned}$$

Given that our interest is to understand the relationships between investments in the *IT Stock* (*T*) and productivity, we now focus on those basis functions in the

Table 7 Basis functions in MARS model

ID	Coefficient	Formula	Expanded formula
BF0	10.522		
BF1	+0.641	Max(0, $\log_e L$ – 16.709)	
BF2	−0.886	Max(0, 16.709 – $\log_e L$)	
BF3	+0.982	(DRG = 0)	
BF4		(DRG = 1)	
BF5	+0.366	Max(0, $\log_e K$ – 13.864)	
BF9	−0.788	Max(0, $\log_e T$ – 14.172) × BF4	Max(0, $\log_e$T – 14.172) × (DRG = 1)
BF10	+0.899	Max(0, 14.172 – $\log_e T$) × BF4	Max(0, 14.172 – $\log_e T$) × (DRG = 1)
BF13		Max(0, $\log_e L$ – 13.473) × BF3	Max(0, $\log_e L$ – 13.473) × (DRG = 0)
BF15	−2.388	Max(0, 13.570 – $\log_e T$) × BF13	Max(0, 13.570 – $\log_e T$) × max(0, $\log_e$L – 13.473) × (DRG = 0)
BF16	−0.160	Max(0, $\log_e K$ – 15.277) × BF9	Max(0, $\log_e$K – 15.277) × max(0, $\log_e$T – 14.172) × (DRG = 1)
BF17	+0.335	Max(0, 15.277 – $\log_e K$) × BF9	Max(0, 15.277 – $\log_e$K) × max(0, $\log_e$T – 14.172) × (DRG = 1)
BF18	+0.171	Max(0, $\log_e L$ – 13.473) × BF9	Max(0, $\log_e$L – 13.473) × max(0, $\log_e$T – 14.172) × (DRG = 1)

regression splines equation that include *T*. Table 8 describes regions of the input variable space, the corresponding "IT Impact Formula" columns (in terms of BFs that include *T*), and "Direction of Impact of Investments in *IT Stock*." The regions are shown using the knots of the relevant basis functions. To obtain the direction of "Impact of Investments in *IT Stock*," we differentiate the relevant "IT Impact Formula" with respect to $\log_e T$.

With regard to the impact of investments in the *IT Stock* on productivity, this regression splines model suggests that

1. Investments in the *IT Stock* have a statistically significant impact on productivity, while under other conditions, the investments in the *IT Stock* do not appear to have a statistically significant impact on productivity (see Table 8).
2. When investments in the *IT Stock* have a statistically significant impact on productivity, the rate of this impact that appears could be positive, negative, or none (see Table 8), contingent on the amounts invested in *Non-IT Labor* (*L*) and/or *Non-IT Capital* (*K*).
3. The impact of investments in the *IT Stock* (*T*) differs when the *DRG* applies (i.e., *DRG* = 1) than when it does not apply (i.e., *DRG* = 0). For example, when the *DRG* applies, investments in the *IT Stock* (*T*) have a statistically significant impact on productivity, contingent on the amounts invested in the *IT Stock* (*T*) and also *Non-IT Labor* (*L*) and/or *Non-IT Capital* (*K*), and this impact is not uniform but could be positive or negative (see Fig. 3). On the other hand, when the *DRG* does not apply, investments in the *IT Stock* (*T*) could have a

Table 8 Rate of impact of investments in the IT Stock on productivity

Region of input variable space			IT impact formula	Impact of investments in IT stock
DRG	$\log_e T$	$\log_e K$		
0	<13.570		$-2.388 \times (13.570 - \log_e T) \times (\log_e L - 13.473)$	*Positive* If $\log_e L > 13.473$
	≥13.570		0	*None*
1	<14.172		$+0.899 \times (14.172 - \log_e T)$	*Negative* $(-0.899 < 0)$
	≥14.172	<15.277	$-0.788 \times (\log_e T - 14.172)$ $+0.335 \times (15.277 - \log_e K) \times (\log_e T - 14.172) + 0.171 \times (\log_e L - 13.473) \times (\log_e T - 14.172)$	*Positive* if $\log_e K < \log_e K_\tau$; *Negative* if $\log_e K > \log_e K_\tau$; where $\log_e K_\tau = 15.277 + (0.171 \times (\log_e L - 13.473) - 0.788)/0.335)$
		>15.277	$-0.788 \times (\log_e T - 14.172) - 0.160 \times (\log_e K - 15.277) \times (\log_e T - 14.172) + 0.171 \times (\log_e L - 13.473) \times (\log_e T - 14.172)$	*Positive* if $\log_e K < \log_e K_\tau$; *Negative* if $\log_e K > \log_e K_\tau$; where $\log_e K_\tau = 15.277 + (0.171 \times (\log_e L - 13.473) - 0.788)/0.160)$

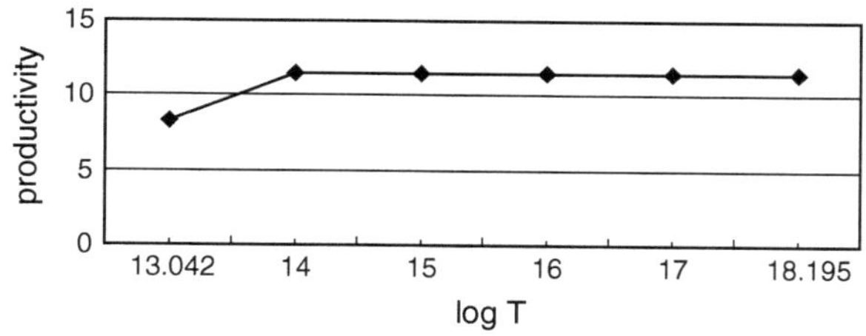

Fig. 2 IT investment and productivity when the DRG does not apply and $\log_e K < \log_e K_{knot1}$ and $\log_e L_{knot1} < \log_e L < \log_e L_{knot2}$, where $\log_e K_{knot1}$ is 13.864, $\log_e L_{knot1}$ is 13.473, and $\log_e L_{knot2}$ is 16.709

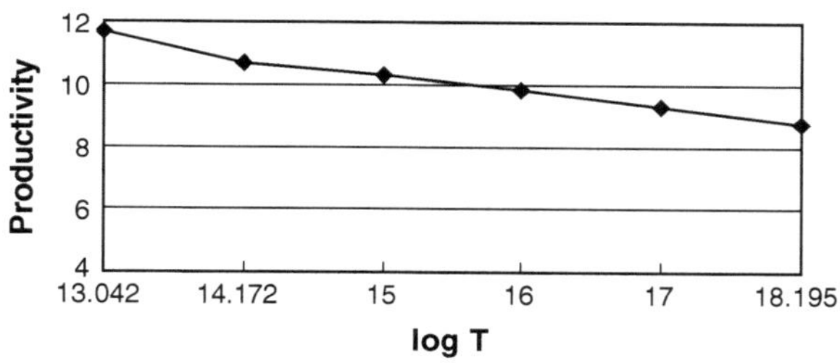

Fig. 3 IT investment and productivity when the DRG applies and $\log_e K > \log_e K_{knot2}$, and $\log_e L_{knot1} < \log_e L < \log_e L_{knot2}$, where $\log_e K_{knot2}$, is 15.277, $\log_e L_{knot1}$ is 13.473, and $\log_e L_{knot2}$ is 16.709

positive statistically significant impact on productivity only when the amount invested in the *IT Stock* does not exceed 13. 570 (the threshold defined by the knot $\log_e T$) (see Fig. 2).

4. With regard to the issue of whether investments in IT should be considered as a substitute to investments in *Non-IT Labor*, our RS results suggest that in general, such a position does not hold.
5. With regard to the complementary impact of investments in *Non-IT Capital* (*K*) on the direction of the impact (i.e., positive or negative) of *IT Stock*, the relevant threshold for *K* is contingent on the amount invested in *L* (e.g., *positive* if $\log_e K < \log_e K_\tau$; *negative* if $\log_e K > \log_e K_\tau$; where $\log_e K_\tau = 15.277 + (0.171 \times (\log_e L - 13.473) - 0.788)/0.335)$ (see Fig. 3, Appendix). Figure 3 shows the IT impact when $\log_e K > \log_e K_\tau$.
6. The overall impact of the investments in *IT Stock* on organizational productivity is not uniform but varies with values of the input variables.

Again, these results suggest that investments in the *IT Stock* can have an impact on productivity and the rate of the impact is in fact contingent on the amounts invested in *Non-IT Labor*, the *IT Stock,* and/or *Non-IT Capital*, as well as to whether the *DRG* applies. Thus, the results also indicate that it is possible for the productivity paradox to occur.

Table 9 Results of regression analysis

Variable	DRG = 0		DRG = 1	
	Coefficients	*P*-value	Coefficients	P-value
$\log_e T$	−0.539	0.581	**1.595**	**0.050**
$\log_e L$	2.402	0.018	−0.248	0.748
$\log_e K$	0.034	0.946	1.246	0.001
$\log_e T \times \log_e L$	**0.184**	**0.001**	**0.551**	**0.000**
$\log_e T \times \log_e K$	**−0.123**	**0.042**	−0.027	0.549
$\log_e L \times \log_e K$	0.114	0.035	−0.096	0.020
$(\log_e T)^2$	−0.070	0.496	**−0.595**	**0.000**
$(\log_e L)^2$	−0.318	0.000	−0.407	0.003
$(\log_e K)^2$	0.005	0.812	0.088	0.002
R^2	**0.951**		**0.941**	

2.4 Results from Regression Analysis

Our regression analysis involved the use of the translog production function. We adopted this theoretical production function because it allows for interaction between the input variables, and as such can be used to explore the complementarity or substitutability of investments in IT with that of the other inputs, particularly *Non-IT Labor*. We did separate regression analysis for the case when *DRG* was not in force (i.e., $DRG = 0$) and for the case when *DRG* was in force (i.e., $DRG = 1$). These results are reported in Table 9. As can be seen, both models have high R-squared values (i.e., 0.951 and 0.941). We then explored the conditions under which investments in IT would have a positive or negative impact on productivity by examining the IT component of the relevant regression equations. These results are reported in Table 10. For the case when *DRG* does not apply (i.e., $DRG = 0$), these results suggest that the direction of the impact of IT investments on productivity is conditioned by the relationship between the amount invested in *Non-IT Labor* (i.e., *T*) and the amount invested in *Non-IT Capital*.

Table 10 Results of regression analysis—direction of impact of IT investments

DRG	IT impact formula	Direction of impact of IT investments
0	$0.184 \times \log_e T \times \log_e L$ $- 0.123 \times \log_e T \times \log_e K$	*Positive* if $\log_e L > (0.123/0.184) \times \log_e K$; *Negative* if $\log_e L < (0.123/0.184) \times \log_e K$;
1	$1.595 \times \log_e T$ $+ 0.551 \times \log_e T \times \log_e L$ $- 0.595 \times (\log_e T)^2$	*Positive* if $\log_e T < (0.551 \times \log_e L + 1.595)/(0.595 \times 2)$; *Negative* if $\log_e T > (0.551 \times \log_e L + 1.595)/(0.595 \times 2)$;

For the case when *DRG* does apply (i.e., $DRG = 1$), these results suggest that the direction of the impact of IT investments on productivity is conditioned by the amount invested in *Non-IT Labor*.

2.5 Comparison of Overall Results from Data Mining Techniques and Regression

In this section, we compare the results of regression with those from both RT and MARS. The results of the previous subsections indicate that each technique provides a model with a high R-squared value, suggesting that each model is strong. As shown in Table 11, both data mining techniques consistently indicate that impact of investments in the *IT Stock* is not uniform but is dependent on the level of other investments as well as whether the *DRG* applies or not. Our results indicate that relationship between investments in the *IT Stock* and *Productivity* differs when the *DRG* applies than when it did not apply. Since *DRG* is a time-based variable, this suggests that the nature of this relationship could be dependent on time in a manner that has not accommodated other variables, such as *Non-IT Labor*, *Non-IT Capital*, and/or *IT Stock*.

A proposition that has been a temptation for top managers with regard to organizational performance is that investments in IT are more effective than investments in *Non-IT Labor*, and so investments in IT should be substituted for investments in *Non-IT Labor*. While the results of our three analytical techniques differ at the level of detail, they all suggest that at a minimum, such a proposition should be treated with caution. Such a cautious approach is consistent with the underlying premise of the contingency theory of organizations which posits that organizational performance is based on the fit between relevant variables (Barua et al. 1996; Selto et al. 1995; Bergeron et al. 2001). With regard to our study, contingency theory suggests that the impact of IT investment on organizational productivity is contingent on one or more complementary factors including whether appropriate levels of investments in non-IT were also made. The results from all three techniques suggest that while investments in the *IT Stock* can have a positive impact on productivity, if the decision is made to continuously increase investments in the *IT Stock* while simultaneously decreasing investments in *Non-IT Labor*, then at some point, the productivity paradox could occur. This result is either expressed directly [e.g., Impact of IT Investment is positive only if $\log_e L > 13.473$ (see Table 8); for $DRG = 1$, the impact of IT Investment is positive only if $\log_e T < (0.551 \times \log_e L + 1.595)/(0.595 \times 2)$ (see Table 10)] or this result could be expressed indirectly [e.g., Impact of IT Investment is positive only if $\log_e L > (0.123/0.184) \times \log_e K$ (see Table 10)]. Our results, therefore, suggest that a complementary relationship exists between investments in the *IT Stock* and investments in *Non-IT Labor*. They also suggest that a complementary relationship exists between investments in the *IT Stock* and investments in

Table 11 Comparison of results from each technique

Issue	RT	MARS	Regression
IT Stock impact	Could be positive or nonexistent, depending on the level of investments in the *IT Stock*, *Non-IT Labor* and *Non-IT Capital*, and also *DRG*	Could be positive, negative, or none depending on the level of investments in the *IT Stock*, *Non-IT Labor* and *Non-IT Capital*, and also *DRG*	Could be positive, negative, or none depending on the level of investments in *Non-IT Labor, Non-IT Capital*, and also *DRG*
	Not uniform	Not uniform	Uniform
Complementary of *Non-IT Labor* (*L*) with the *IT Stock* (T)	Yes	Yes	Yes
Effect of time period (DRG)	Yes	Yes	Yes
IT productivity paradox	Possible to occur	Possible to occur	Possible to occur
R^2	0.829	0.945	0.941 for DRG = 0, 0.951 for DRG = 1

Non-IT Capital [e.g., for *DRG* = 0, the impact of IT investment is positive only if $\log_e L > (0.123/0.184) \times \log_e K$ (see Table 10)].

Although all three analytic techniques can provide insight into understanding the complex relationship between investments in the *IT Stock* and productivity, it appears to us that the data mining approaches offer some advantages over regression including

1. Regression analysis requires the assumption of a theoretical production function (i.e., translog in this paper) and each assumed production function imposes restrictions on the relationships between variables that can be induced from the data. On the other hand, both data mining approaches do not require any such assumption but rather use the data to determine the appropriate functional form of the production function.
2. The data mining techniques can induce cutoff points for investments in the *IT Stock* since any investments beyond the cutoff point will be counterproductive [e.g., impact of investments in the *IT Stock* is negative if $\log_e T > 13.570$ (see Table 8)].
3. Compared to regression analysis that induces a single global model [e.g., a single causal relationship for each *DRG* value (see Table 10)], which applies to the entire predictor space, the data mining techniques induce several local models that are tuned to the characteristics of distinct regions of the predictor space [e.g., several causal relationships for each *DRG* value (see Table 8)].

2.6 Discussion on Business Meaning of Results

So what is the business meaning of our results? Our data are at the level of granularity of financial amounts invested in IT (Administrative and Medical, Capital and Labor), *Non-IT Labor*, and *Non-IT Capital* rather than at a finer level of granularity (e.g., specific IT capital acquisitions such as ERP or BI software, types of Non-IT professionals, customer service strategy). Thus, our results can only speak to providing guidance for top-level budgeting decisions with regard to amounts invested in IT, *Non-IT Labor*, and *Non-IT Capital*. For example, the budgeting process in many healthcare organizations involves top-down budgeting, bottom-up budgeting, or a hybrid of these two approaches. In top-down budgeting, top management develops and approves the enterprise-wide budget even before project and department budgets are developed. Middle- and lower-level managers then break down these estimates into project and departmental budgets. Our regression splines model (see Tables 7 and 8) could be used to provide guidelines for the top-down budgeting process and to do "What-If" analysis. Our results suggest that if *DRG* applies, then the amount invested in the *IT Stock* has to exceed a threshold before it will have a positive impact on productivity. If the amount invested in the *IT Stock* exceeds this threshold, then for these investments to have a positive impact on productivity, it suggests that the amount invested in *Non-IT Capital*

Table 12 Impacts of Administrative IT for Selected Scenarios

Scenario			Impact Formula for Administrative IT	Direction of Rate of Impact
DRG	log_c**AIT**	log_c**K**		
0	$\leq$ 15.767	$\leq$ 10.0 10	- 0.006*(13.593 - log_cK)*(15.767- log_cAIT)	**Positive** 0.006*(13.593 - log_cK) > 0
		(10.010, 13.593)	- 0.006*(13.593 - log_cK) *(15.767- log_cAIT) - 0.004*(16.005- log_cAIT)*(log_cK - 10.010)	**Positive** (0.006*(13.593 - log_cK) + 0.004*(1og$_c$K - 1 0.010)) > 0
		> 13.593	+ 0.050*(log_cK- 13.593)*(15.767 -log_cAIT) - 0.004 *(16.005- log_cAIT)* (log_cK - 10.010)	**Positive** if $log_cK > log_cK_{\tau1}$; **Negative** if log$_c$l<. < $log_cK_{\tau1}$ Where $log_cK_{\tau1}$= 13.90
	(15.767, 16.005)	$\leq$ 10.010	0	**None**
		> 10.010	- 0.004*(16.005 - log_cAIT)*(log_cK – 10.010)	**Positive** (+ 0.004* (log_cK – 10.010) > 0)
	> 16.005		0	**None**
1	$\leq$ 13.987	$\leq$ 10.010	+ 0.077 *(13.987 - log_cAIT)	**Negative** (-0.077)
		(10.010, 13.593)	+ 0.077 *(13.987 - log_cAIT) - 0.004*(16.005 - log_cAIT)*(log_cK - 10.010)	**Positive** if $log_cK > log_cK_{\tau1}$; **Negative** if $log_cK < log_cK_{\tau2}$ where $log_cK_{\tau2}$ = 10.010 + (0.077/0.004) = 29.26
		> 13.593	+ 0.077 *(13.987- log_cAIT) + 0.050*(log_cK - 13.593)*(15.767 - log_cAIT) - 0.004*(16.005 - log_cAIT)*(log_cK - 10.010)	**Positive** if log$_c$K<. > $log_cK_{\tau3}$; **Negative** if $log_cK < log_cK_{\tau3}$ Where $log_cK_{\tau3}$= 12.2306

cannot exceed a threshold that is partly determined by the amount invested in *Non-IT Labor* (see Table 8).

It is often assumed that labor productivity can be improved in three ways: increasing the level of capital applied per unit of labor, improving in the quality of labor by appropriate additional education/training of the current workforce and/or acquisition of appropriately trained or trainable employees, and increase in multifactor productivity (MFP) such as improvement in the production methods of business processes or in the quality of the firms output. While several studies suggest that the *IT Stock* can be a net substitute for *Non-IT Labor*, our results suggest that this holds only under certain conditions (see Table 8). One reason for this result is that IT use is often associated with an improvement in the quality (e.g., skill level) of *Non-IT Labor*, which may result in an increase in the cost of *Non-IT Labor* as in general, higher skilled workers earn higher wages (Autor et al. 1998). To the extent that investments in *Non-IT Labor* can be considered to be representative of the quality of *Non-IT Labor*, and investment in *Non-IT Capital* can be considered to be partially representative of the level of *Non-IT Capital* per unit of *Non-IT Labor*, our results suggest that improvement in productivity will not automatically follow from simply increasing the amount invested in these two areas and also the *IT Stock*. Our results for the productivity of *Non-IT Labor* (see Table 12) also suggest Administrative IT can lead to improvements in labor productivity but only under certain conditions (see Table 12). Although the level of granularity of our data does not allow us to determine how IT was used (e.g., automate critical business processes, and/or provide value-enhancing information in an appropriate and timely manner, and/or transform the production methods of business processes), our results (see Tables 7, 8 and 12) could be interpreted as suggesting that it is only appropriate multifactor calibration of the amounts invested in IT, *Non-IT Labor*, and *Non-IT Capital* that will result in improvement in labor productivity and overall production.

3 Conclusion

In this paper, we revisited the relationship between IT and productivity at the firm level. In addition to the traditional regression analysis approach, we utilized data mining techniques in order to open the "black box" and identify conditions, including interactions with other predictors, under which IT could have an impact on productivity. Our study shows that investments in IT have a statistically significant impact on productivity and that this impact could be positive, negative, or none depending on the conditions, such as amounts invested in *Non-IT Labor* and/or *Non-IT Capital*. In addition, this impact of IT differs for the period before the *DRG* implementation and the one after the *DRG* implementation. A possible reason for this difference is the changes in hospital operations associated with the *DRG* implementation.

Our study made several contributions to the IT and productivity research literature. First, it exposes the fact that the relationship between IT and productivity is very complex including being nonlinear and conditional. Second, the impact of investments in *IT Stock* on productivity occurs within the context of conditional complementary relationship with investments in *Non-IT Labor* (*L*) and/or *Non-IT Capital* (*K*). To have greater impact on productivity, the levels of non-IT investments along with IT Investments should be considered altogether. Third, our research also suggests that the IT productivity paradox could still occur.

In this study, we investigated what has happened in the past between IT investment and productivity. For future research, we would like to build a model based on the datasets including organizations selected as the innovative IT users in the *Information Week 500* and the ones that are in similar size and industry but not selected as the innovative IT users. Using a data mining approach, we could investigate factors that could make organizations efficient or inefficient, when considering IT investment. Also, we could compare the thresholds of IT investment to determine whether these two groups of organizations have different thresholds. If different, we can investigate the degree of difference and the rates of IT impact in each group. This might assist inefficient hospitals in attaining the productivity levels of efficient hospitals.

Acknowledgments Material in the chapter previously has appeared in "Reexamining the Impact of Information Technology Investment on Productivity Using Regression Tree and Multivariate Adaptive Regression Splines," in the *Information Technology & Management (9:4*, 285–299 (2008)).

Appendix: Derivatives of IT Impact Formulas

In this appendix, we display the results (Table 13) of the differentiating with respect to $\log_e T$, each "IT Impact Formula" in Table 8.

Table 13 Derivatives of IT impact formulas

IT impact formula	Derivative with respect to $\log_e T$
$-2.388 \times (13.570 - \log_e T) \times (\log_e L - 13.473)$	$-2.388 \times (\log_e L - 13.473) \times (-1)$ $= 2.388 \times (\log_e L - 13.473) > 0$
$+ 0.899 \times (14.172 - \log_e T)$	$0.899 \times (-1) = -0.899 < 0$
$-0.788 \times (\log_e T - 14.172)$ $+ 0.335 \times (15.277 - \log_e K) \times$ $(\log_e T - 14.172)$ $+ 0.171 \times (\log_e L - 13.473) \times$ $(\log_e T - 14.172)$	$-0.788 \times 1 + 0.335 \times (15.277 - \log_e K) \times 1$ $+ 0.171 \times (\log_e L - 13.473) \times 1$ $= 0.335 \times (15.277 - \log_e K) +$ $(0.171 \times (\log_e L - 13.473) - 0.788)$ This derivative is *positive* if: $0.335 \times (15.277 - \log_e K) +$ $(0.171 \times (\log_e L - 13.473) - 0.788) > 0$ $\Rightarrow (0.171 \times (\log_e L - 13.473) - 0.788) > - 0.335 \times (15.277 - \log_e K)$ $\Rightarrow (0.171 \times (\log_e L - 13.473) - 0.788)/0.335 > -15.277 + \log_e K$ $\Rightarrow 15.277 + (0.171 \times (\log_e L - 13.473) - 0.788)/0.335 > \log_e K$ Let $\log_e K_\tau = 15.277 + (0.171 \times (\log_e L - 13.473) - 0.788)/0.335$ • This derivative is *positive* if $\log_e K < \log_e K_\tau$ • This derivative is *negative* if $\log_e K > \log_e K_\tau$
$-0.788 \times (\log_e T - 14.172)$ $- 0.160 \times (\log_e K - 15.277) \times$ $(\log_e T - 14.172)$ $+ 0.171 \times (\log_e L - 13.473) \times$ $(\log_e T - 14.172)$	-0.788×1 $- 0.160 \times (\log_e K - 15.277) \times 1$ $+ 0.171 \times (\log_e L - 13.473) \times 1$ $= -0.160 \times (\log_e K - 15.277) +$ $(0.171 \times (\log_e L - 13.473) - 0.788)$ This derivative is *positive* if: $- 0.160 \times (\log_e K - 15.277) +$ $(0.171 \times (\log_e L - 13.473) - 0.788) > 0$ $\Rightarrow (0.171 \times (\log_e L - 13.473) - 0.788) > 0.160 \times (\log_e K - 15.277)$ $\Rightarrow (0.171 \times (\log_e L - 13.473) - 0.788)/0.160 > (\log_e K - 15.277)$ $\Rightarrow 15.277 + (0.171 \times (\log_e L - 13.473) - 0.788)/0.160 > \log_e K$ Let $\log_e K_\tau = 15.277 + (0.171 \times (\log_e L - 13.473) - 0.788)/0.160$ • This derivative is *positive* if $\log_e K < \log_e K_\tau$ • This derivative is *negative* if $\log_e K > \log_e K_\tau$

References

Autor D, Katz L, Krueger A (1998) Computing inequality: have computers changed the labor market? Q J Econ 113(4):1169–1213

Barua A, Sophie Lee CH, Whinston AB (1996) The calculus of reengineering. Inf Syst Res 7(4):409–428

Bergeron F, Raymond L, Rivard S (2001) Fit in strategic information technology management research: an empirical comparison of perspectives. Omega 29:125–142

Brynjolfsson E, Hitt LM (1996) Paradox lost? Firm-level evidence on the returns to information systems spending. Manage Sci 42(4):541–558

Deichmann J, Eshghi A, Jaigjtpm D, Sayek S, Teebagy N (2002) Application of multiple adaptive regression splines (MARS) in direct response modeling. J Interact Mark Autumn, 15–27

Dewan S, Min CK (1997) The substitution of information technology for other factors of production: a firm level analysis. Manage Sci 43(12):1660–1675

Groff ER, Wartell J, McEwen JT (2001) An exploratory analysis of homicides in Washington, DC. In: The 2001 American society of criminology conference

Hitt LM, Brynjolfsson E (1996) Productivity, business profitability, and consumer surplus: three different measures of information technology value. MIS Q 20(2):121–142

Kudyba S, Diwan R (2002) Research report: increasing returns to information technology. Inf Syst Res 13(1):104–111

Lichtenberg FR (1995) The output contributions of computer equipment and personnel: a firm-level analysis. Econ Inf New Technol 3(4):201–217

Menon NM, Lee B (2000) Cost control and production performance enhancement by IT investment and regulation changes: evidence from the healthcare industry. Decis Support Syst 30(2):153–169

Menon NM, Lee B, Eldenburg L (2000) Productivity of information systems in the healthcare industry. Inf Syst Res 11:83–92

Mukopadhyay T, Lerch F, Mangal V (1997) Assessing the impact of information technology on labor productivity—a field study. Decis Support Syst 19:109–122

Selto FH, Renner CJ, Mark Young S (1995) Assessing the organizational fit of a just-in-time manufacturing system: testing selection, interaction and systems models of contingency theory. Acc Organ Soc 20(7–8):665–684

Shao B, Lin W (2001) Measuring the value of information technology in technical efficiency with stochastic production frontiers. Inf Softw Technol 43:447–456

Shin H (2006) The impact of information technology on the financial performance of diversified firms. Decis Support Syst 41:698–707

Chapter 10
Overview on Cluster Analysis

Kweku-Muata Osei-Bryson and Sergey Samoilenko

This chapter provides an overview of cluster analysis. Its main purpose is to introduce the reader to the major concepts underlying this data mining (DM) technique, particularly those that are relevant to the chapter that involves the use of this technique. It also provides an illustrative example of cluster analysis.

1 Introduction

Clustering attempts to automatically partition a dataset into a meaningful set of mutually exclusive clusters (or segments). While segmentation can be done directly by humans without the use of DM software, DM-based clustering attempts to segment the data into natural clusters that are relatively homogenous with respect to some similarity metric (e.g., Euclidean distance). The characteristics of the resulting clusters are thus not subjectively determined. However, in some cases, the resulting clusters may have no natural meaning to the user.

There are different reasons for doing clustering, two major categories of which are as follows:

1. Finding a set of natural clusters and the corresponding description of each cluster. This is relevant if there is the belief that there are natural groupings in the data. In some cases, such as fraud detection, there is interest in finding segmentations that include outlier clusters (e.g., Aggarwal and Yu 2001). For some

K.-M. Osei-Bryson (✉)
Department of Information Systems, Virginia Commonwealth University,
301 W. Main Street, Richmond, VA 23284, USA
e-mail: KMOsei@VCU.Edu

S. Samoilenko
Department of Computer Science, Averett University, 420 W Main Street Danville,
VA 24541, USA
e-mail: SSamoilenko@Averett.Edu

K.-M. Osei-Bryson and O. Ngwenyama (eds.), *Advances in Research Methods for Information Systems Research*, Integrated Series in Information Systems 34,
DOI: 10.1007/978-1-4614-9463-8_10,

other cases, there may be a preference in finding a segmentation that includes a pair of clusters that provides the lowest mean and highest mean for each variable. And still for other cases, there may be a preference for finding segmentations that include certain specified variables as important discriminating variables between the clusters. For each of these cases, it is possible that multiple segmentations could apply. Thus, it might not be appropriate to use a single clustering algorithm and/or parameter setting to find the appropriate set of "natural" clusters.

2. Improving the performance of predictive modeling and other DM techniques when there are many competing patterns in the data. For this case, there is an interest in obtaining a segmentation that will result in an improvement in performance and also offer a convenient description of the clusters so that it will facilitate the assignment of new observations to the appropriate predictive model. However, it is possible that multiple segmentations could apply.

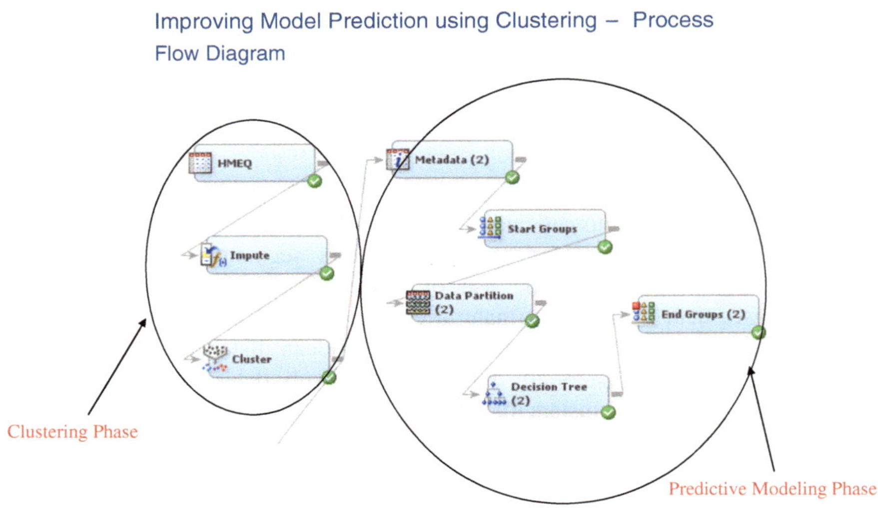

In addition to its use by practitioners, in recent years, clustering has also been applied as a research tool for exploring important research problems in information systems and other areas (Balijepally et al. 2011). For example, Rai et al. (2006) used cluster analysis to explore various questions including whether the assimilation of electronic procurement innovations increases procurement productivity; Okazaki (2006) used it to characterize mobile internet adopters, and Wallace et al. (2004) used it to characterize software project risks. For these applications of clustering, the given researcher was interested in obtaining "natural" grouping and corresponding descriptions. Typically these research projects did not involve the exploration of multiple clustering algorithms and parameter settings but involved the use of default parameter settings. As noted by Balijepally (2006):

> A vast majority of IS studies have however neither reported the algorithm used in the study nor the distance measure used, though some improvement has been noticed over the two time periods. Non-reporting of such basic requirements of cluster analysis leads

to suspicion that researchers could be blindly using the default settings in the computer packages without a clear understanding of the methodology or the implications of the decision choices involved therein.

Yet it is known that for a given clustering algorithm, different parameter settings could result in different segmentations, several of which could be relevant to the given research question(s).

2 Understanding the Output of Clustering

Different approaches may be used to understand the output of clustering including the following:

1. Building a decision tree (DT) with the cluster label as the target variable and using the associated rules to conveniently describe each cluster (e.g. Fig. 1) as well as explain how to assign new records to the correct cluster (e.g., Mathers and Choi 2004).
2. Examining the distribution of variable values from cluster to cluster. (e.g. Fig. 2). Typically this involves the domain expert(s) doing comparison of cluster means for the relevant variables (e.g., Wallace et al. 2004).
3. Visual inspection by a domain expert of a graphical 2-dimensional (2-D) representation of the clustering output in order to assess the validity of the results (e.g., Bittman and Gelbrand 2009; Kimani et al. 2004).
4. A hybrid of the approaches above.

3 Clustering Algorithms

There are numerous algorithms available for doing clustering. They may be categorized in various ways such as hierarchical (e.g., Murtagh 1983; Ward 1963) or partitional (e.g., Chen et al. 2004; Mc Queen 1967), deterministic or probabilistic (e.g., Bock 1996), and hard or fuzzy (e.g., Bezdek 1981; Dave 1992). Hierarchical methods generate a hierarchy of clusters from the given dataset using either a top-down or bottom-up iterative process. Partitional methods (e.g., *k*-Means, *k*-Median) divide the given dataset into a user-specified number of clusters (Figs. 1, 2).

_k_-Means: Divides the dataset into _k_ clusters

Step 1: Pick *k* seed points as the initial clusters' CENTROIDs
Step 2: Assign each object (i.e., data unit) to the cluster whose CENTROID is closest to the given object
Step 3: Let the New CENTROID of each cluster be the MEAN of objects in the cluster
Step 4: If the new CENTROID of each cluster is the same as the old CENTROID or is sufficiently close, then TERMINATE; otherwise, REPEAT Steps 2 through 4
The final Set of Clusters generated by *k*-Means is sensitive to the choice of initial cluster Centroids

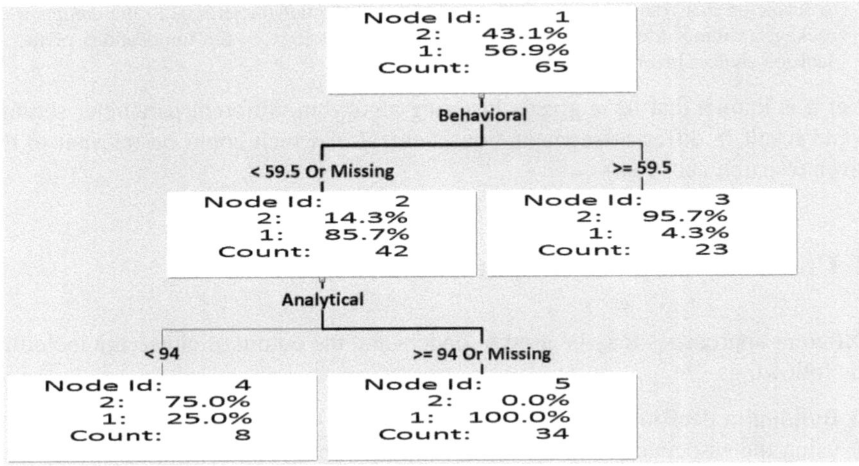

Node = 3	IF **Behavioral** >= 59.5 THEN Number of Observations = 23 Predicted: _ **SEGMENT_=2 = 0.96** Predicted: _SEGMENT_=1 = 0.04
Node = 4	IF **Behavioral** <59.5 AND **Analytic** < 94 THEN Number of Observations = 8 Predicted: **_SEGMENT_=2 = 0.75** Predicted: _SEGMENT_=1 = 0.25
Node = 5	IF **Behavioral** < 59.5 AND **Analytic** >= 94 THEN Number of Observations = 34 Predicted: _SEGMENT_=2 = 0.00 Predicted: **_SEGMENT_=1 = 1.00**

Fig. 1 Example of a DT that describes a segmentation with 2 clusters (i.e., "1", "2"). Input variables: *Analytical*, *Behavioral*, *Conceptual*, *Directive*

Hierarchical methods may be categorized as being agglomerative or divisive, with the former involving combining clusters generated in previous iterations using some combination rule and the latter involving dividing clusters generated in previous iterations.

Agglomerative	• A bottom-up approach that merges pairs of clusters • Starts with each data point in its own cluster • At each step, merges the closest pair of clusters
Divisive	• Top-down approach involving binary division of clusters • Start with all data points in an all-inclusive cluster • At each step, split a cluster until each data point is in its own cluster

Cluster Means

Cluster	Analytical	Behavioral	Conceptual	Directive
1	105.46	43.81	73.03	75.49
2	83.00	69.82	75.04	72.14
Overall: Mean	95.78	55.02	73.89	74.05

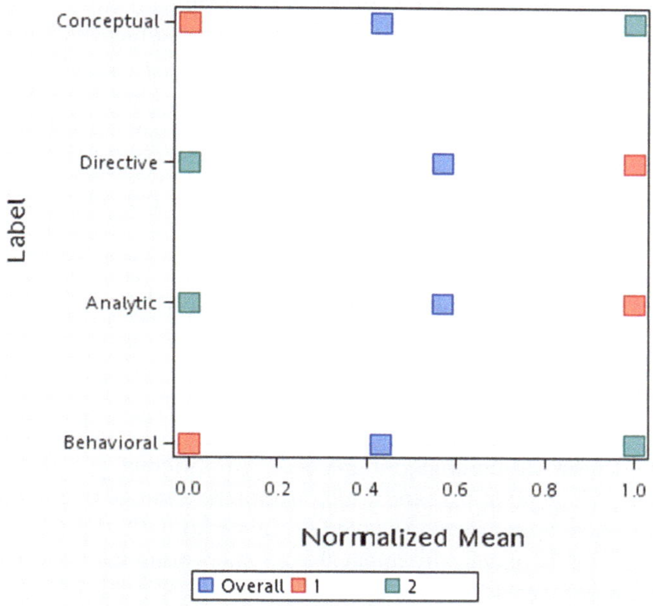

Relative Importance of Variables with regards to Differentiating between the Clusters

Variable	Analytical	Behavioral	Conceptual	Directive
Relative Weight	0.946	1.000	0.000	0.711

Fig. 2 Example of cluster means that describes a segmentation with 2 clusters. Input variables: *Analytical*, *Behavioral*, *Conceptual*, *Directive*

Example of an agglomerative algorithm: average-link algorithm for generating *g* clusters from *m* data points:

Step 1: Assign each data point to its own cluster, so that there are *m* clusters. A partitional method could be used to do this assignment

Step 2: Merge the most similar pair of clusters into a single cluster. The result is that there is now one less cluster

Step 3: Compute the distance between the new cluster and each of the old clusters

Step 4: Repeat steps 2 and 3 until there are *g* clusters

Similarity Metrics

Distances are normally used to measure the similarity or dissimilarity between two data points. Commonly used metrics include the *Minkowski distance*, cosine, and correlation. The general form of the *Minkowski distance* is

$$d(i,j) = \sqrt[q]{\left(\left|x_{i1} - x_{j1}\right|^q + \left|x_{i2} - x_{j2}\right|^q + \cdots + \left|x_{\mathrm{ip}} - x_{\mathrm{jp}}\right|^q\right)}$$

where $i = (x_{i1}, x_{i2}, \ldots, x_{\mathrm{ip}})$ and $j = (x_{j1}, x_{j2}, \ldots, x_{\mathrm{jp}})$ are two p-dimensional data objects, and q is a positive integer. Examples for various values of q are as follows:

$q = 1$: Mean absolute deviation	$q = 2$: Euclidean distance
• Distance$(x, y) = \Sigma_i \lvert x_i - y_i \rvert$	• Distance$(x, y) = (\Sigma_i (x_i - y_i)^2)^{1/2}$
• Resistant to outliers	• Sensitive to outliers
• Centroid value is the median	• Centroid value is the arithmetic mean
q > 1 & < 2	**q > 2**
• Distance$(x, y) = (\Sigma_i (x_i - y_i)^q)^{1/q}$	• Distance$(x, y) = (\Sigma_i (x_i - y_i)^q)^{1/q}$
• Resistant to outliers	• Sensitive to outliers

4 Evaluating the Output of Clustering Algorithms

Typically, for a given dataset, different algorithms may give different sets of clusters, so it is never clear which algorithm and which parameter settings (e.g., number of clusters) are the most appropriate. As noted by Jain et al. (1999): "There is no clustering technique that is universally applicable in uncovering the variety of structures present in multidimensional data sets." They thus raised the following questions: "How is the output of a clustering algorithm evaluated? What characterizes a 'good' clustering result and a 'poor' one?" Ankerst et al. (1999) also commented that "Most of the recent research related to the task of clustering has been directed toward efficiency. The more serious problem, however, is effectivity, i.e., the quality or usefulness of the result."

4.1 The Issue of Quality: Assessing Cluster Validity

Jain et al. (1988) describe cluster validity as the assessment of the set of clusters that is generated by the given clustering algorithm. They note that there are three approaches for assessing validity:

1. External assessment, which involves comparing the generated segmentation (i.e., set of clusters) with an a priori structure, typically provided by some domain experts.
2. Internal assessment, which attempts to determine whether the generated set of clusters is "intrinsically appropriate" for the data. Several techniques have

been proposed for internal assessment of cluster validity (e.g., Bezdek 1981; Dunn 1974; Gordon 1999; Kaufman and Rousseeuw 1990; Osei-Bryson 2005; Ramze Rezaee et al. 1998; Tibshirani and Walther 2005). A central idea behind most of these approaches is that a valid, high-quality segmentation should consist of clusters that are both cohesive and well separated, although some techniques give greater emphasis to one of these properties.
3. Relative assessment, which involves comparing two segmentations (i.e., 2 sets of clusters) based on some performance measures (e.g., Dubes 1983; Jain and Dubes 1988) and measure their relative performance.

4.2 The Issue of Usefulness: Goals for Clustering

There are several possible goals for a clustering exercise, including the following:

1. *Segmentation should include (exclude) outliers*: This problem has several applications including fraud detection (e.g., Aggarwal and Yu 2001).
2. *Segmentation should include a pair of clusters that provides the lowest and highest means for each variable*: An example of this problem is a theory-building study (Wallace et al. 2004) where the interest was in finding clusters that provide the characteristics that could be used to describe low-, medium-, and high-risk projects. In this case, a pair of clusters was found that provided the smallest and largest means for almost all the variables.
3. *Segmentation should include user-specified variables as important discriminating variables*: There are several reasons why this goal may be important to users. For example, as noted by Huang et al. (2005) "It is well-known that an interesting clustering structure usually occurs in a subspace defined by a subset of the initially selected variables. To find the clustering structure, it is important to identify the subset of variables." This could also be relevant for theory-building exercises. Balijepally (2006) suggested that "Studies where the clustering variables are tightly linked to theory are considered deductive. If the variables are generated based on expert opinion, the approach is deemed cognitive … Legitimacy accorded to the pursuit of theories from reference disciplines in IS research could be one reason working in favor of adopting a deductive approach…. One suggestion for improvement would be to consider using a cognitive approach to variable selection over a pure inductive approach. This involves tapping expert opinions either from other IS researchers, IS practitioners, or both."

5 Illustrative Example

This clustering exercise focuses on a set of 18 transition economies (TEs) that spans the period from 1993 through 2002 that was used in previous study of Samoilenko and Osei-Bryson (2010). The reason for doing cluster analysis was to inquire whether or not all 18 TEs are similar with regard to their investments in and revenues from ICT or whether there are multiple naturally occurring groups. This set of TEs consists only of former members of the Soviet bloc that started the process of transition at approximately the same time and includes Albania, Armenia, Azerbaijan, Belarus, Bulgaria, Czech Republic, Estonia, Hungary, Kazakhstan, Kyrgyz Republic, Latvia, Lithuania, Moldova, Poland, Romania, Slovak Republic, Slovenia, and Ukraine. We were interested in a segmentation (i.e., set of clusters) that did not involve any outlier cluster, where a cluster was considered to be an outlier cluster if it contained less than 10 % of the data points in the dataset.

The original data were obtained from the following:

- the *WDI* database (web.worldbank.org/WBSITE/EXTERNAL/DATASTATISTICS) and
- the Yearbook of Statistics (2004) (www.itu.int/ITU-D/ict/publications) *of International Telecommunication Union* (www.itu.int).

Each row in the dataset is identified by Country Name and Year. For cluster analysis, we used the following eight variables:

Revenue	**Investments**
• Total telecom services revenue (% of GDP in current US$)	• Annual telecom investment per capita (current US$)
• Total telecom services revenue per capita (current US$)	• Annual telecom investment (% of GDP in current US$)
• Total telecom services revenue per worker (current US$)	• Annual telecom investment per worker (current US$)
• Total telecom services revenue per telecom worker (current US$)	• Annual telecom investment per telecom worker (current US$)

SAS Enterprise Miner (EM) was used to perform cluster analysis of the data set. The variables that we used are not measured on the same scale, so prior to cluster analysis, we transformed the data by standardizing the variables. We generated segmentations with 5, 4, 3, and 2 clusters. We present details for the 5- and 4-cluster segmentations, followed by a summary of the results (Table 1).

Table 1 Summary statistics

Number of clusters	Number of data points in each cluster
5	(C1: 10), (C2: 11), (C3: 20), (C4: 3), (C5: 136)
4	(C1: C1: 10) (C2: 32), (C3: 3), (C4: 135)
3	(C1: 30), (C2: 3), (C3: 147)
2	(C1: 72), (C2: 108)

Outlier clusters are underlined

5-Cluster Segmentation

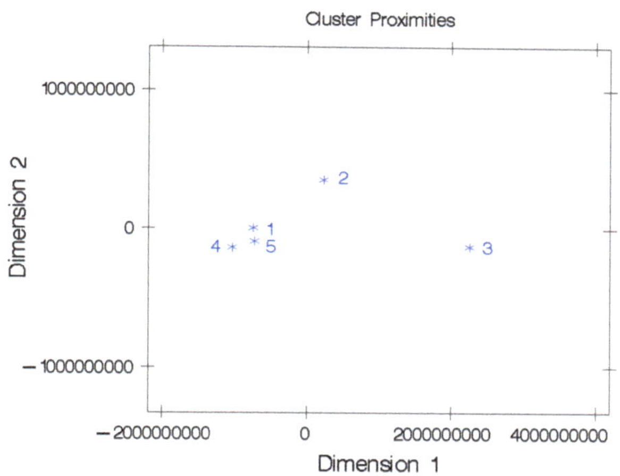

Name	Importance	Measurement	Type	Label
TOTAL_TELECOM_SERVICES_REVENUE_C	0	interval	num	Total telecom services revenue(Current US $)
TOTAL_TELECOM_SERVICES_REVENUE_0	0	interval	num	Total telecom services revenue(Current US $ per per
TOTAL_TELECOM_SERVICES_REVENUE_1	1	interval	num	Total telecom services revenue(Current US $ per wor
ANNUAL_TELECOM_INVESTMENT_CURREN	0	interval	num	Annual telecom investment(Current US $)
ANNUAL_TELECOM_INVESTMENT_CURRE0	0	interval	num	Annual telecom investment(Current US $ per person)
ANNUAL_TELECOM_INVESTMENT____OF_	0.7506022753	interval	num	Annual telecom investment(% of GDP)
ANNUAL_TELECOM_INVESTMENT_CURRE1	0.95367343	interval	num	Annual telecom investment(Current US $ per worker)
ANNUAL_TELECOM_INVESTMENT_CURRE2	0	interval	num	Annual telecom investment(Current US $ per telecom
TOTAL_TELECOM_SERVICES_REVENUE_2	0	interval	num	Total telecom services revenue(Current US $ per te
PRODUCTIVITY_RATIO_PER_TELECOM_W	0.3814699143	interval	num	Productivity ratio per telecom worker (revenue/inve
FULL_TIME_TELECOMMUNICATION_STAF	0	interval	num	Full-time telecommunication staff(% of total labor

CLUSTER	Frequency of Cluster	Root-Mean-Square Standard Deviation	Maximum Distance from Cluster Seed	Nearest Cluster	Distance to Nearest Cluster	Total telecom services revenue(Current U
1	10	0.6910952169	3.3838196355	5	2.8685729343	301079379.57
2	11	0.9125016132	3.6211143581	3	4.3069207936	1167300217.7
3	20	0.9272675431	4.5743045211	2	4.3069207936	3247370982
4	3	0.79872865	3.039460512	5	7.3359108746	52491522.551
5	136	0.5521570041	3.5951652596	1	2.8685729343	343798794.92

Three of the clusters (C1, C4, C5) are close together, while clusters C2 and C3 are situated further away. Clusters C1, C2, and C4 are outlier clusters as the number of observations (frequency of cluster) in each is below the 10 % threshold (i.e., 18 data points).

4-Cluster Segmentation

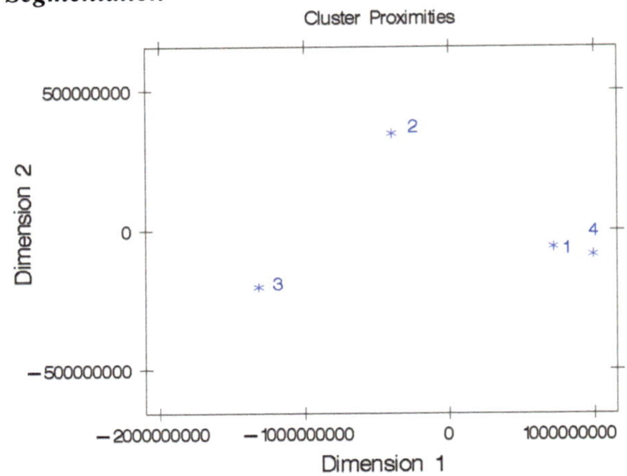

Name	Importance	Measurement	Type	Label
TOTAL_TELECOM_SERVICES_REVENUE_C	0	interval	num	Total telecom services revenue(Current US $)
TOTAL_TELECOM_SERVICES_REVENUE_0	0	interval	num	Total telecom services revenue(Current US $ per per
TOTAL_TELECOM_SERVICES_REVENUE_1	1	interval	num	Total telecom services revenue(Current US $ per wor
ANNUAL_TELECOM_INVESTMENT_CURREN	0.2894552211	interval	num	Annual telecom investment(Current US $)
ANNUAL_TELECOM_INVESTMENT_CURRE0	0	interval	num	Annual telecom investment(Current US $ per person)
ANNUAL_TELECOM_INVESTMENT____OF_	0	interval	num	Annual telecom investment(% of GDP)
ANNUAL_TELECOM_INVESTMENT_CURRE1	0.6168238947	interval	num	Annual telecom investment(Current US $ per worker)
ANNUAL_TELECOM_INVESTMENT_CURRE2	0	interval	num	Annual telecom investment(Current US $ per telecom
TOTAL_TELECOM_SERVICES_REVENUE_2	0	interval	num	Total telecom services revenue(Current US $ per te
PRODUCTIVITY_RATIO_PER_TELECOM_W	0.2844069536	interval	num	Productivity ratio per telecom worker (revenue/inve
FULL_TIME_TELECOMMUNICATION_STAF	0	interval	num	Full-time telecommunication staff(% of total labor

CLUSTER	Frequency of Cluster	Root-Mean-Square Standard Deviation	Maximum Distance from Cluster Seed	Nearest Cluster	Distance to Nearest Cluster	Total telecom services revenue(Current U
1	135	0.549179016	5.8872861968	3	4.510702041	308195775.29
2	10	0.8792442624	3.5227988604	3	4.5254221857	1251400954.6
3	32	0.9412023199	5.5697550356	1	4.510702041	2294834772.9
4	3	0.79872865	3.039460512	1	7.3562791919	52491522.551

Clusters C1 and C4 are close together (accounting for 138 data points out of 180) with the clusters C3 and C2 being fairly removed. Clusters C2 and C4 are outlier clusters as the number of observations in each is below the 10 % threshold (i.e., 18 data points).

Table 2 Contents of the 2-cluster segmentation (1993–2002)

Contents of the cluster 1	Contents of the cluster 2
Albania (1993–2002)	**Czech Republic (1993–2002)**
Armenia (1993–2002)	Estonia (1994–2002)
Azerbaijan (1993–2002)	**Hungary (1993–2002)**
Belarus (1993–2002)	Bulgaria (2002)
Bulgaria(1993–2001)	Latvia (1994, 1995, 1997–2002)
Slovak Republic (1993, 1994, 1999)	Lithuania (1999–2002)
Kazakhstan (1993–2002)	**Slovenia (1993–2002)**
Kyrgyz Republic (1993–2002)	**Poland (1993–2002)**
Latvia (1993, 1996)	Slovak Republic (1995–1998, 2000–2002)
Lithuania (1993–1998)	
Moldova (1993–2002)	
Romania (1993–2002)	
Ukraine (1993–2001)	

BOLD indicates that the given country is in the same cluster for the entire 1993-2002 period

The reader may recall that our aim was to generate a segmentation that did not involve any outlier clusters. Of the four segmentations that were generated, only the 2-cluster segmentation met this requirement. The contents of the associated clusters are displayed in Table 2, where countries whose data points are completely contained in a single cluster are in bold.

By using cluster analysis, we were able to come up with a solution that partitions our dataset into two clusters. The issue of the validity of this segmentation could be addressed using an external evaluation approach to cluster validity, where a domain expert's opinion can provide external confirmation of the validity of this segmentation. Such domain expert support for this 2-cluster segmentation is provided by Piatkowski (2003), who concluded that in the period "between 1995 and 2000 ICT capital has most potently contributed to output growth in the Czech Republic, Hungary, Poland, and Slovenia." Thus, it could be suggested that we were able to separate 18 TEs into the two groups, one group of TEs which benefits the most from the investments in telecom and another group where the benefits are less pronounced.

References

Aggarwal CC, Yu PS (2001) Outlier detection for high dimensional data. In: Proceedings of the 2001 ACM SIGMOD international conference on management of data, pp 37–46

Ankerst M, Breunig M, Kriegel H-P, Sander J (1999) OPTICS: ordering points to identify the clustering structure. In: Proceedings of ACM SIGMOD'99 international conference on the management of data, Philadelphia, PA, 1999, pp 49–60

Balijepally V, Mangalaraj G, Iyengar K (2011) Are we wielding this hammer correctly? A reflective review of the application of cluster analysis in information systems research. J Assoc Inf Syst 12(5):375–413

Balijepally V (2006) Application of cluster analysis in information systems research: a review. A & M University, Prairie View

Bittman A, Gelbrand R (2009) Visualization of multi-algorithm clustering for better economic decisions—the case of car pricing. Decis Support Syst 47(1):42–50

Bezdek J (1981) Pattern recognition with fuzzy objective function algorithms. Plenum Press, New York

Bock H (1996) Probability models in partitional cluster analysis. Comput Stat Data Anal 23:5–28

Chen S-C, Ching R, Lin Y-S (2004) An extended study of the k-means algorithm for data clustering and its applications. J Oper Res Soc 55:976–987

Dave R (1992) Generalized fuzzy C-shells clustering and detection of circular and elliptic boundaries. Pattern Recogn 25:713–722

Dubes R (1983) Cluster analysis and related issues. In: Chen C, Pau L, Wang P (eds) Handbook of pattern recognition and computer vision. World Scientific Publishing Co. Inc., River Edge, pp 3–32

Dunn J (1974) Well separated clusters and optimal fuzzy partitions. J Cybern 4:95–104

Gordon A (1999) Classification. Chapman & Hall, New York

Kaufman L, Rousseeuw P (1990) Finding groups in data. Wiley, New York

Huang J, Ng M, Rong H, Li Z (2005) Automated variable weighting in k-means type clustering. IEEE Trans Pattern Anal Mach Intell 27(5):657–668

Jain A, Dubes R (1988) Algorithms for clustering data. Prentice-Hall advanced reference series. Prentice-Hall Inc., Upper Saddle River

Jain A, Murty M, Flynn P (1999) Data clustering: a review. ACM Comput Surv 31(3):264–323

Kimani S, Lodi S, Catarci T, Santucci G, Sartori Vidamine C (2004) A visual data mining environment. J Visual Lang Comput 15:37–67
Mathers W, Choi D (2004) Cluster analysis of patients with ocular surface disease blepharitis, and dry eye. Arch. Ophthalmol. 122:1700–1704
Mc Queen J (1967) Some methods for classification and analysis of multivariate observations. In: Lecam LM, Neyman J (eds) Proceedings of the 5th Berkeley symposium on mathematical statistics and probability. California Press, Berkeley, pp 281–297
Murtagh F (1983) A survey of recent advances in hierarchical clustering algorithms which use cluster centers. Comput J 26:354–359
Okazaki S (2006) What do we know about mobile internet adopters? A cluster analysis. Inf Manage 43(2):127–141
Osei-Bryson K-M (2005) Assessing cluster quality using multiple measures. The next wave in computing, optimization, and decision technologies, pp 371–384
Rai A, Tang X, Brown P, Keil M (2006) Assimilation patterns in the use of electronic procurement innovations: a cluster analysis. Inf Manage 43(3):336–349
Ramze Rezaee M, Lelieveldt B, Reiber J (1998) A new cluster validity index for the fuzzy c-mean. Pattern Recogn Lett 19:237–246
Samoilenko S, Osei-Bryson K-M (2010) Determining sources of relative inefficiency in heterogeneous samples: methodology using cluster analysis, DEA and neural networks. Eur J Oper Res 206(2):479–487
Tibshirani R, Walther G (2005) Cluster validation by prediction strength. J Comput Graph Stat 14(5):11–28
Wallace L, Keil M, Rai A (2004) Understanding software project risk: a cluster analysis. Inf Manage 42:115–155
Ward J (1963) Hierarchical grouping to optimize an objective function. J Am Stat Assoc 58:236–244

Chapter 11
Overview on Data Envelopment Analysis

Sergey Samoilenko

The chapter provides a general introductory overview of data envelopment analysis. Its main purpose is to introduce the reader to the major concepts underlying this nonparametric technique. After familiarizing the reader with the general process used in calculating the scores of relative efficiency, the chapter presents an overview of various orientations and types of DEA models. In conclusion, the chapter gives an overview of using DEA for the purposes of constructing Malmquist index, a popular tool for measuring changes in efficiency over time; a brief example is used to illustrate major points.

1 Introduction

Data envelopment analysis (DEA) is a nonparametric method of measuring the efficiency of decision-making units (DMU). Any collection of similar entities could comprise a set of DMUs and be subjected to DEA, as long as the chosen entities transform the *same type* of inputs into the *same type* of outputs. Inputs and outputs, taken together, constitute a common *DEA model* for all DMUs in the sample. Thus, for all intents and purposes of DEA, every DMU in the sample is represented completely by the values of its inputs and outputs of the DEA model. Because some of the inputs or outputs of the DEA model could be more significant than others, DEA offers a decision maker a flexibility of assigning various weights to the inputs and outputs of the model; the equal weighting is commonly utilized as a default.

The empirical foundation of DEA eliminates the need for some of the assumptions and limitations of traditional efficiency measurement approaches. As a result,

S. Samoilenko (✉)
Department of Computer Science, Averett University, 420 W Main St,
Danville, VA 24541, USA
e-mail: SSamoilenko@Averett.Edu

K.-M. Osei-Bryson and O. Ngwenyama (eds.), *Advances in Research Methods for Information Systems Research*, Integrated Series in Information Systems 34,
DOI: 10.1007/978-1-4614-9463-8_11,

DEA could be used in cases where the relationships between the multiple inputs and multiple outputs of the DEA model are complex or unknown. Consequently, a DEA model is not necessarily comprised of the real inputs that are converted into the real outputs as it is implied by a production process. Rather, a DEA model is better perceived as a collection of the inputs that are in some way or form important to the outputs of the transformation process under an investigation of a decision maker.

2 The General Idea Behind the Approach

The original DEA model was introduced in 1978 by Charnes, Cooper, and Rhodes, and it is commonly called the **CCR Model** (an abbreviation consisting of first letters of the authors' names). This model allowed representing multiple inputs and outputs of each DMU as a single abstract "meta-input" and a single "meta-output." Consequently, the efficiency of each DMU could be represented as a ratio of the abstract input to the abstract output, and the resulting efficiency value could then be used for comparison with other DMUs in the set. Using the techniques of *Linear Programming* (LP), this comparison results in efficiency ranking of each DMU in the given set, where the highest ranking DMU is considered to be 100 % efficient and is assigned a perfect score of "1". Because multiple DMUs could receive the same score, there could be multiple efficient DMUs in the given set. As a result, DEA *envelops* the data set with the boundary points represented by the efficient DMUs—by connecting the boundary points, an investigator could obtain a visual representation of the *efficient frontier* for a given set of DMUs.

In the case of a *non-relaxed* LP (where values of the inputs and outputs are restricted to integers), a score of less than "1" means that some other DMUs in the sample could produce the given level of outputs using less inputs (e.g., in *Output-Oriented* model, see below) or could utilize the given level of the inputs more efficiently by producing higher level of the outputs (in the case of *Input-Oriented* model). In the case of the *relaxed* LP (when the integer constraint is relaxed by allowing the inputs and outputs to take interval values), however, a DMU receiving a score of less than "1" could still be considered *weakly efficient*. This weak efficiency, signified by the presence of *non-zero slacks* (the presence of some unutilized amount of inputs or outputs), takes place in the case if there is no other DMU which is better given the levels of every input or output of the model.

One of the benefits of DEA is that it allows for comparing relative efficiencies of DMUs in the absence of monetary information regarding the inputs and outputs. Let us consider the simple scenario when we need to compare a group of firms that produce the same type of retail merchandise. For all intents and purposes, every firm could be viewed as an input–output system that receives raw materials and produces merchandise. If the cost of the input, e.g., raw material, and the cost of the output, e.g., price received for the merchandise, is known, then the firms could be compared via a simple output–input ratio, where a firm with the greatest ratio is the most efficient in the group. Unfortunately, this simple scenario could not be successfully utilized in modern business environment for a few reasons. First, globalization impacted a business world via information transparency in such way, that a competition based on the low cost of inputs and high price of outputs is no longer feasible. Thus, firms are forced to compete on the basis of internal capabilities—processes that convert inputs into outputs. Under such circumstances in order to evaluate the relative efficiency of a firm, many different types of inputs and outputs need to be considered. Second, monetary information regarding the cost of inputs and the price of the outputs is rarely available; furthermore, not all inputs and outputs could be monetized.

The “size” of DMUs is important for determining the efficiency of DMUs in regard to its scale; if a given DMU considered to be efficient to the point where any changes in its size—the values of inputs and outputs—will result in the decrease in efficiency, then the DMU is labeled as *scale efficient*. Because it is reasonable to expect a great variety of the DMUs in the sample in regard to their size, it is also possible that smaller DMUs will be compared to larger ones—this will result in comparison of differently scaled DMUs. A concept of *technical efficiency* is introduced to counter the impact of scale heterogeneity via the assumption that the DMUs are of the same scale in the sense that no DMU is operating beyond its optimal capacity. An important and useful measure of an overall or *allocative* efficiency is applied when the inputs and outputs could be monetized—in those situations when costs and prices for the inputs and outputs of the DMU model are available to the investigator. It is worth noting that the estimates of allocative efficiency of DMUs, which is also referred to as *price efficiency*, might not agree with estimates of technical efficiency. To illustrate the difference between the two consider an example when a company A is deemed efficient in terms of its utilization of resources (e.g., it is technically efficient), but it is not efficient in utilization of the costs incurred by the acquisition of the resources (e.g., it is not allocatively efficient). Simply put, while technical efficiency concentrates on utilization of resources, allocative efficiency focuses on costs and prices—think about the project management environment, where a project manager may be concerned with the wise allocation of various types of resources, while an accounting manager may prefer to concentrate on whether the dollar amounts associated with the cost of acquisition of the resources are utilized the most efficiently.

3 Major Assumptions

One of the fundamental assumptions of DEA is that of a *functional similarity* of the DMUs in a sample. This simply means that in order to compare, meaningfully, the relative efficiencies of DMUs in the sample, these DMUs must be similar in terms of utilization of the inputs and production of the outputs. The assumption of functional similarity is enforced via the specification of the common DEA model that prescribes the same set of the inputs and the same set of the outputs for all DMUs in the sample. Thus, a set of DMUs could be represented by a group of basketball teams, firms, countries, hospitals, as long as all DMUs in the sample are defined by the same DEA model.

Another assumption of DEA is that of *semantic similarity* of DMUs in the sample, where DEA requires us to make sure that we compare apples and apples, and not apples and oranges; while we could conduct DEA using the data set representing hospitals, we should not apply DEA to the data set representing hospitals and basketball teams. The assumption of semantic similarity is not explicitly enforced within DEA, but rests under the purview of the decision maker.

Yet another assumption supporting DEA is that of the availability of resources, where it is assumed that all DMUs in the sample have approximately the same level of access to the available resources and technology. Under this assumption, an investigator could subject a group of high school basketball teams to DEA, but an inclusion of a few of NBA teams in the sample would clearly violate this assumption.

Finally, an assumption of similarity of the competitive environment requires consideration of the possible differences that DMUs in the sample could face via their local environments—it is expected that DMUs in the sample operate in the similar competitive environment. Under the circumstances when a competitive environment clearly impacts the performance of DMUs, it is often dealt with via inclusion of *environmental variable*. For example, when DEA must be performed on a set of retail stores that operate in clearly different competitive environments, the environmental variable *Market_Competition* could be added to DEA model, where the domain of values of the variable could be "1"—least competitive, "2"—average, and "3"—most competitive.

4 Orientations

One of the benefits of DEA lies in its flexibility, for an investigator could take advantage of several models and orientations that this method has to offer. Thus, for example, the choice of a given DEA model would depend on the underlying economic assumptions about the returns to scale of the process that transforms the inputs into the outputs. Consequently, the different assumptions would yield the different models. As a result, instead of forcing a single perspective, DEA offers multiple vantage points on the process of input–output transformation in the form of the several available to the researcher models and orientations.

The three commonly mentioned orientations of DEA model are input, output, and base oriented. An *Input-Oriented Model* is concerned with the minimization of the use of inputs for achieving a given level of outputs when inputs are controllable. For example, if an investigator wants to identify a most efficient college student among the group of "A" students, then an input-oriented model will fit fine for the purpose, for such inputs as "number of absences" and "number of study hours per week" are clearly under the control of the students—DMUs in the sample. In the case of an input-oriented model, no DMU would be considered efficient if it is possible to decrease any of its inputs without affecting any other inputs and without decreasing level of the outputs.

An *Output-Oriented DEA model,* on the other hand, would be concerned with the maximization of the level of the outputs per given level of the inputs. Thus, it deals with the efficiency of the output production where outputs are controllable. Let us consider an example of investigating relative efficiencies of similar assembly lines, where each line is represented by a fixed number of full-time workers. Under this scenario, the inputs might be considered fixed—an assembly line is limited to the existing number of full-time workers, as well as it is limited to the number of hours workers are allowed to work with no overtime paid. This example lends itself well to being investigated via output-oriented model, for only the outputs of the model could be controlled by each assembly line, while the values of inputs remain unchanged. In the case of an output-oriented model, a DMU could not be considered efficient if it is possible to increase the level of any of its outputs without affecting other outputs or increasing level of any of its inputs.

A *Base-Oriented Model*, unlike the first two, has a dual orientation and would be concerned with the optimal combination of the inputs and outputs. Therefore, this type of DEA model deals with the efficiency of the input utilization and efficiency of the output production, having control over both inputs and outputs within the model. Let us consider comparing multiple construction teams in terms of their relative efficiency. Based on this scenario, every team could control its inputs—the number of people working and the quantity of rented equipment, as well as its outputs—progress of the project. This sort of a situation will be best analyzed via base-oriented DEA model, for inputs and outputs are not fixed.

5 Types of Models

The concepts and methodologies of DEA are now incorporated into four models: CCR, BCC, additive, and multiplicative.

- One of the basic, most straightforward DEA models is CCR, the original model of Charnes, Cooper, and Rhodes. The *CCR* ratio model yields an objective evaluation of overall efficiency and identifies the sources and estimates the amounts of thus identified inefficiencies. The most appropriate case for using CCR is when an investigator has a reason to believe that all DMUs function under the condition of constant return to scale (e.g., when the impact of doubling the inputs of a firm results in the consequent doubling of the outputs).
- The *BCC* (Banker, Charnes, and Cooper) model distinguishes between technical and scale inefficiencies by estimating pure technical efficiency at the given scale of operation and identifying whether increasing, decreasing, or constant returns to scale possibilities are present for further exploitation. The main difference between CCR model and the model of BCC lies in how the returns to scale are handled. While the CCR model assumes constant return to scale, the BCC model is more flexible in that it allows for the variable returns to scale.

Hence, a given DMU is considered to be efficient by CCR model only if it is both scale and technically efficient, while for the same DMU to be considered efficient by BCC model it must only be technically efficient. Thus, if a DMU is considered to be efficient by the CCR model, it will also be considered as such by the BCC model, while the reverse not necessarily being true. In addition to the CCR model, BCC makes the efficiency of a given DMU to be contingent on two conditions: the value of the proportional reduction being equal to one and all slacks being equal to zero. Because this condition implies a zero distance between the DMU and the efficiency frontier, all the sources of inefficiency stem from the presence of the slacks and value of output–input ratio being less than one. Similarly to the CCR model, for a given DMU to be qualified as efficient under the input-oriented model of the BCC, the DMU must simultaneously qualify as efficient under the output-oriented model and vice versa. Figure 1 illustrates the differences in the enveloping surfaces produced by the CCR and BCC models for the set of 7 DMUs with a single input and a single output.

In Fig. 1, the dashed line represents the frontier of the CCR model, with constant returns to scale, while the solid line depicts the enveloping surface produced by the BCC model with the variable returns to scale. As we could see, the CCR model produces an efficiency frontier defined by the single DMU (P2), while the efficiency frontier of the BCC model is produced by the four DMUs (P1, P2, P3, and P4). The DMUs that do not belong to this envelopment surface (or efficient frontier) are operating inefficiently. The most appropriate case for using the BCC model is when a decision maker encounters a context of production growth or decline (e.g., when the impact of doubling the inputs of a firm results in the consequent disproportional increase or decrease in outputs), such as the introduction of new products, or a production environment that functions above the level of its optimum capacity.

- The *Additive* model and its variants do not prioritize the orientation (e.g., input vs. output) of the analysis. Instead of dealing with the problems of minimization of the inputs or maximization of the outputs, the additive models estimate

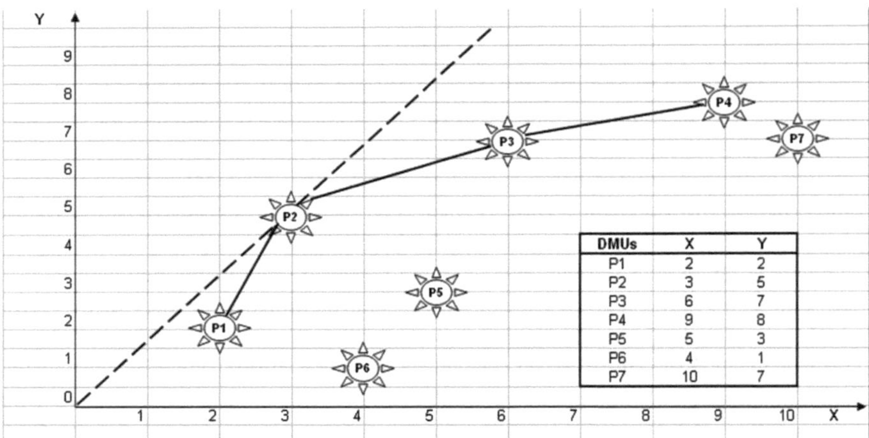

Fig. 1 DEA: CCR (*Dashed line*) and BCC (*Solid line*) models

relative efficiency via *simultaneous* minimization of the inputs *and* maximization of the outputs. Furthermore, while the CCR and BCC rely on *radial* measure of efficiency—proportional minimization of the inputs or maximization of the outputs, additive models rely on *nonradial* measure of efficiency, where individual inputs and outputs are allowed to change at different rates. It seems reasonable to suggest that the use of the additive model will be applicable in the contexts where a decision maker may want to investigate the relative efficiency of the general process of transformation of inputs into outputs. Consider a scenario of an IS consultancy or a marketing agency; under this scenario, it is of value to determine the sources of inefficiencies of an average project team, thus identifying a general composition (in terms of the type and qualification of the members of the team, average types and amounts of technical resources available to the team, etc.) of the team with regard to the general type of the outcomes (e.g., types of deliverables, quality benchmarks). Clearly, some of the input resources, as well as output-related deliverables, will be more important than others (e.g., office supplies vs. laptops), and additive model will allow, via disproportional change of individual inputs and outputs, to arrive at the most beneficial resource to the decision maker configuration of the team.

- The *Multiplicative* model provides a log-linear envelopment of the production frontier. Unlike other DEA models, which rely on additive aggregation of inputs and outputs to calculate the scores of the relative efficiency, multiplicative models rely on multiplication. While the CCR and the BCC DEA models rely on a qualitative grouping of the return-to-scale options into constant, decreasing, or increasing categories, multiplicative models allow for a more precise, quantitative estimation of the return to scale. Albeit rarely applied in practice, the use of multiplicative models is warranted if an investigator suspects that the shape of the enveloped surface is not uniformly convex, but consists of concave and convex regions. An intuitive context for applying the multiplicative model is a production environment, where, for example, each unit of manufacturing equipment (input A in a DEA model) may be associated with the utilization of multiple types of resources—e.g., number of employees per shift (input B), number of working hours per employee per shift (input C), electricity consumption per hour (input D). Under such circumstances, it is only reasonable to utilize the multiplicative aggregation, for the overall input resource (e.g., meta-input) is better expressed via multiplying the number of units by the number of the factors required to operate it during the production.

6 Malmquist Index and Total Factor Productivity

While the benefit of being able to investigate the relative efficiency of DMUs via DEA is undeniable, the investigator may also be interested in *changes* in the scores of the relative efficiency that took place over a period of time. It is only reasonable that the results of DEA will be incorporated into some sort of a policy with the goal of improving the current level of efficiency of the conversion of inputs into outputs. By obtaining the information regarding the changes in the scores over time, a decision maker could evaluate whether a given DMU increased (e.g., positive change in the score), decreased (e.g., negative change in the score), or kept unchanged (e.g., no change in the score) its relative efficiency as compared to other DMUs in the sample. Malmquist index (MI) constructed via DEA offers an investigator such a tool for identifying changes in relative efficiency of the DMUs over time. The following introduction to MI will rely on bringing together the related concepts of *productivity* and *efficiency*, which, while being different measures of performance, become equivalent under the assumption of constant return to scale.

It is commonly assumed that economic growth could be determined by two factors. The first factor, resource accumulation, could lead to high rates of growth, albeit, due to the law of diminishing return, only for a limited period of time. Thus, it is the second factor, growth in productivity that is assumed to allow for attaining sustained economic growth. The productivity is commonly referred to as total factor productivity (TFP) and its growth is measured by the use of Malmquist index.

Based on the idea of the productivity index, originally suggested by Malmquist (1953), Caves, Christensen, and Diewert (1982) defined the Malmquist index of TFP growth. Later, Färe, Grosskopf, Norris, and Zhang (1994) showed that the Malmquist index could be constructed based on the results of DEA. Essentially, the approach is based on performing DEA analysis in two points in time; let us say *T*1 and *T*2. Let us again consider running the example based on the set of 7 DMUs, each with a single input and a single output. Figure 2 shows constructed efficiency frontier at the time *T*1 and, designated by a dashed line, inefficiency of the DMU P5 relative to the frontier.

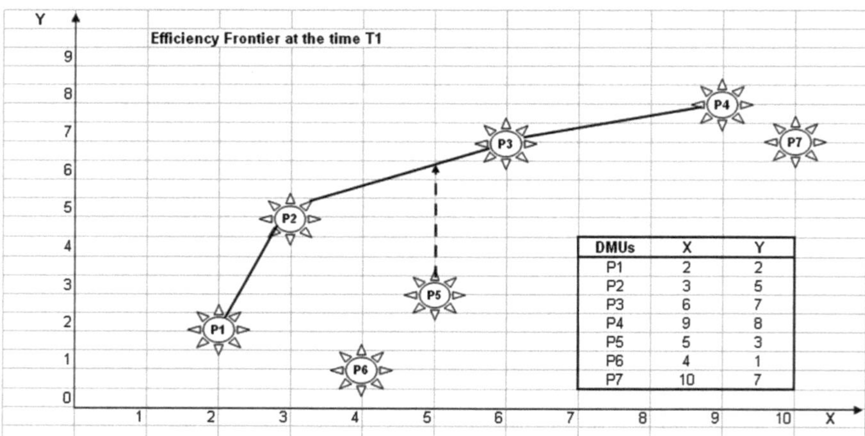

DMUs	X	Y
P1	2	2
P2	3	5
P3	6	7
P4	9	8
P5	5	3
P6	4	1
P7	10	7

Fig. 2 Efficiency frontier, efficient and inefficient DMUs

Then, for a given DMU, the period of time ($t2 - t1$) could be represented by the distance between the data point at the time $t1$ and the data point at the time $t2$. For each DMU, the distance between these data points would be reflective of the change in this DMU's TFP, which is, of course, represented by the Malmquist index.

In the case of economic growth, we would expect that the efficiency frontier for a given set of DMUs would change its position over a period of time. Let us suppose that three DMUs, *P2*, *P3*, and *P5*, have changed their position over the period of time $T2 - T1$ (as depicted in the Fig. 2). Such change is reflected by the new positions of these DMUs, as well as the new position of the efficiency frontier (represented by dashed line). As a result, for a given period in time, change in the position of each DMU could be perceived as consisting of the two components. The first component is the change in distance between a given DMU and the efficient frontier, which reflects the changes in *technical efficiency*, and the second is the change in position of the efficient frontier itself, reflective of the *technological change* (TC) that took place over ($t2 - t1$) period of time. Figure 3 depicts such changes in the frontier, as well as the change in the position of the DMU *P5* relative to the frontier. In this case, change in the position of DMU *P5* relative to the frontier, less the change in the position of the frontier itself, would be represented by a particular value of the Malmquist index for DMU P5.

Thus, a conceptual mechanism of the process of estimating TFP via DEA is straightforward—if the position of the efficient frontier identified by DEA changes over time, the change can be measured by means of MI and decomposed into two components. The first component reflects *Changes in Efficiency* (EC) and is depicted as a change in distance between the position of a given DMU and the efficient frontier. The second component reflects TC and is captured as a change in position of the efficient frontier itself over the period of time. Grifell-Tatjé and Lovell (1995) demonstrated that value of MI could be greater, equal to, or less than 1, reflecting, respectively, growth in productivity, stagnation, or decline in productivity.

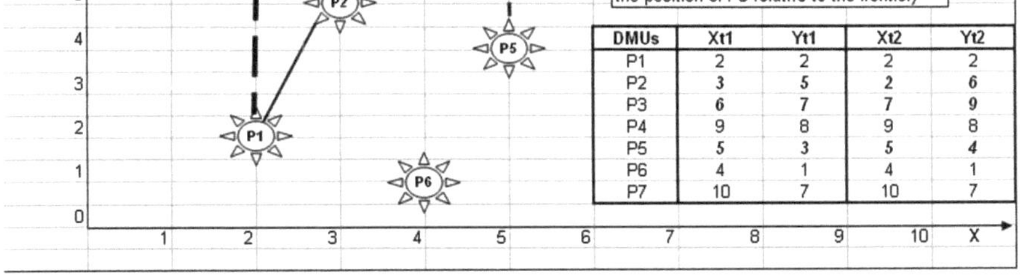

DMUs	Xt1	Yt1	Xt2	Yt2
P1	2	2	2	2
P2	3	5	2	6
P3	6	7	7	9
P4	9	8	9	8
P5	5	3	5	4
P6	4	1	4	1
P7	10	7	10	7

Fig. 3 Components of Malmquist index: TC and EC

However, given that MI is composed of two components, EC and TC, the condition for growth in productivity is that the *sum* of EC and TC is greater than 1 (Grifell-Tatjé and Lovell 1995; Grifell-Tatjé and Lovell 1999; Trueblood and Coggins 2003). Perhaps, a simple example could be offered to illustrate the issues associated with the composite nature of growth in productivity. Let us consider an office worker whose productivity was measured via a single input variable—number of hours worked, and a single output variable—number of reports prepared. Let us also suppose that the worker A received a new workstation at the beginning of the year. Results of DEA indicate that the office worker *A* exhibited an overall growth in productivity, relative to office worker *B*, during the period "before new workstation—after new workstation"—prior to the new workstation the worker *A* was able to produce 20 reports per 8 h, and since the workstation was installed the worker *A* is able to produce 22 reports per 8 h. Any manager will be interested in keeping the productivity growth going; however, the question is *what the driver of growth is*?

It is possible that the office worker A became more productive solely on the basis of a better technology (e.g., the new workstation has a faster processor, more RAM, better display)—this is the growth via TC component. It is also possible that growth in productivity has nothing to do with the new workstation and took place because of the increased efficiency of the office worker (via experience, know-how, training, etc.)—this is the growth via EC component.

This brings about five possible scenarios of growth in productivity that we consider in order.

1. First, MI > 1 when EC > 1 and TC > 1. In this case, the growth in productivity is balanced, and the office worker *A* was able to use better technology *while* becoming more efficient. The office worker *A* is ahead of the technology frontier. There is an indication that the available technology is utilized effectively and efficiently.
2. Second, MI > 1, when EC > 1 and TC = 1. This means that the office worker *A* did become more efficient, but the upgrade to the new workstation did not cause any relative changes in efficiency—as in the scenario where the office worker *B* also got a new workstation. In this case, the investments in a new workstation were made just to keep up with the advancements in technology. The office worker A keeps up with the constantly advancing technology frontier. There is an indication that the skills of the worker *A* are ahead of the available technology, but the level of the technology is adequate relative to what is available to the worker *B*.
3. Third, MI > 1, when EC > 1, and TC < 1. This means that despite getting a new workstation, and becoming more efficient, the worker *A* did not receive a workstation comparable in quality with the one received by the worker *B*. The office worker *A* falls behind the advancing technology frontier.
4. Fourth, MI > 1, when EC = 1, and TC > 1. This means that the office worker *A* did not change her relative level of efficiency at all; however, the overall growth in productivity was observed due to the advanced technology available to *A*—the technology is the driver of the growth.

5. Fifth, MI > 1, when EC < 1, and TC > 1. In this scenario, worker A actually became less relatively efficient, but her productivity is saved by the superior technology (e.g., the workstation available to worker B is less technologically advanced than the one given to worker A). The efficiency of worker A has decreased, but, overall, the productivity growth was obtained due to the superior technology.

The simple examples above offer an important insight—a rational decision-making process associated with the goal of increasing productivity must rely on the information regarding the sources of growth in productivity. Clearly, it is not wise to invest in the latest-greatest technology if the users cannot take advantage of it, as well as it is not the best idea to invest in the work force development if the available technology is not adequate.

Appendix

DEA model types

Type	Charnes-Cooper transformation	LP dual ("Farrell model")	LP dual solution (Score)
Input-oriented	$\max z = \sum^{s} \mu_r y_{r0}$ Subject to $\sum^{s} \mu_r y_{r0} - \sum^{m} v_i x_{ij} \leq 0$ $\sum^{m} v_i x_{ij} \leq 1$ $\mu_r, v_i \geq 0$	$\Theta^* = \min \Theta$ Subject to $\sum^{n} x_{ij}\lambda_j \leq \Theta x_{10} \quad i = 1, 2, \ldots, m;$ $\sum^{n} y_{rj}\lambda_j \geq y_{r0} \quad r = 1, 2, \ldots, s;$ $\lambda_j \geq 0 \quad j = 1, 2, \ldots, n;$	*Solution*: $\Theta^* \leq 1$; *Score*: If $\Theta^* < 1$, DMU is inefficient; If $\Theta^* = 1$, DMU is efficient
Output oriented	$\min q = \sum^{m} v_i x_{i0}$ Subject to $\sum^{m} v_i x_{ij} - \sum^{s} \mu_r y_{rj} \geq 0$ $\sum^{s} \mu_r y_{r0} = 1$ $\mu_r, v_i \geq \varepsilon$	$\sum^{j} z_j x_{jn} \geq \Theta u_{jm} \quad m = 1, 2, \ldots, M;$ $\sum^{j} z_j x_{jn} \leq x_{jn} \quad n = 1, 2, \ldots, N;$ $z_j \geq 0 \quad j = 1, 2, \ldots, J;$	*Solution:* $\Theta^* \geq 1$; *Score:* If $\Theta^* > 1$, DMU is inefficient; If $\Theta^* = 1$, DMU is efficient

Adapted from Cooper et al. (2004)

References

Caves DW, Christensen LR, Diewert WE (1982) The Economic Theory of Index Numbers and the Measurement of Input, Output, and Productivity. Econometrica 50:1393–1414

Cooper WW, Seiford LM, Zhu J (2004) Data envelopment analysis: history, models and interpretations. In: Cooper WW, Seiford LM, Zhu J (eds) Handbook on data envelopment analysis, Chap. 1. Kluwer Academic Publishers, Boston, pp 1–39

Färe R, Grosskopf S, Norris M, Zhang Z (1994) Productivity Growth, Technological Progress, and Efficiency in Industrialized Countries. Am Econ Rev 84:374–380
Grifell-Tatjé C, CAK Lovell (1995) A Note on the Malmquist productivity index. Econ Lett 47:169–175
Grifell-Tatjé C, CAK Lovell (1999) Profits and Productivity. Manage Sci, INFORMS 45(9):1177–1193
Malmquist S (1953) Index numbers and indifference surfaces. Trabajos de Estatistica 4:209–242
Seol H, Choi J, Park G, Park Y (2007) A framework for benchmarking service process using data envelopment analysis and decision tree. Expert Syst Appl 32(2):432–440
Trueblood MA, Coggins J (2003) Intercountry agricultural efficiency and productivity: A Malmquist index Approach. World Bank, Washington, D.C

Chapter 12
ICT Infrastructure Expansion in Sub-Saharan Africa: An Analysis of Six West African Countries from 1995 to 2002

Felix Bollou

The World Bank, International Monetary Fund, the UN and International Telecommunications Union (ITU) argue that ICT infrastructure and informatization are prerequisites to adequate development in the present era. The ITU has proposed a framework for measuring ICT efficiency and its impact on social development. However, it offered no advice on how to develop, model and implement this proposal. In this chapter, we use data envelopment analysis to address one aspect of the ITU proposal, the measurement of the efficiency of investments in ICT infrastructure development. The study makes two important contributions: (1) It provides a methodology for assessing the efficiency of investments in ICT; (2) it provides insights into the structuring development policies to benefit from effective allocation of scarce resources in developing countries.

1 Introduction

Since 1995, many African nations have been increasing investment in ICT infrastructure in response to business and social demands and influence from international development organizations. Scholars have suggested that these investments in ICT will make significant contributions to social and economic development by fostering 'opportunities of the global digital economy' to their communities (UNDP 2001). Braga et al. (2000) suggest that ICTs will help Africa 'leap frog' the stages of economic development. These opinions of the potential of ICT to transform the economic and social development of Africa are not universal. Some scholars argue that less developed countries (LDCs), unlike developed countries have little of the

F. Bollou (✉)
School of Information Technology and Communications, American University of Nigeria, Yola 640001, Adamawa, Nigeria
e-mail: bollou@gmail.com

K.-M. Osei-Bryson and O. Ngwenyama (eds.), *Advances in Research Methods for Information Systems Research*, Integrated Series in Information Systems 34,
DOI: 10.1007/978-1-4614-9463-8_12,

supporting infrastructure that is necessary to capitalize on the productive capacity of ICT (Landauer 1995). Others argue that it is difficult to provide evidence of the impact of ICT on social and economic development because of the time lag between investment and productivity results (Avgerou 1998). Nonetheless, the UN ICT task force has advised governments of LDCs to prioritize and focus on ICT infrastructure expansion as an integral part of their poverty eradication strategies. But the range of developmental challenges faced by African policy makers necessitates prudence in the allocation of scarce resources. Are ICT infrastructures providing an adequate return to warrant continued heavy investment in the face of health pandemics and increasing needs for other civil infrastructure? Empirical evidence on the impact on ICT investment would be useful to policy analysts in deciding this question and to decide on what level of future ICT investments is appropriate within their development strategies. However, there is still limited empirical research in this area (Akpan 2000; Mwesige 2004).

The objective of this short paper is to report empirical research on ICT expansion in sub-Saharan Africa. The findings reported here are an analysis of statistical data on the ICT sectors of six African countries, namely Benin, Burkina Faso, Cameroon,[1] Cote d'Ivoire, Mali, and Senegal. The analysis focuses on two main questions: (1) Are investments in ICT technical efficient with regard to the building and expansion of ICT infrastructure? (2) Are investments in ICT resulting in revenue growth and contributing to growth in GDP (a component of development). To analyze the relative efficiency of these countries, we use data envelopment analysis, a well-known and widely used method for evaluating economic units, such as countries, local governments, and industries (Shafer and Byrd 2000; Wang et al. 1997; Chen and Zhu 2004; Färe et al. 1983).

2 Background on the Countries

All six of the countries in the study are considered LDCs and placed close to the bottom of UN Human Development Index (HDI) rankings in 2002. Two of these countries, Cote d'Ivoire and Cameroon have the largest populations of the group, 16.4 and 17.1 millions, respectively, and are demographically more similar than the rest. Cameroon and Cote d'Ivoire also have the highest literacy rate of the group, 75 and 59.8 %, respectively. Both countries have fairly large urban populations, comprehensive universities and a very high level of enrollment in primary and secondary education. Senegal which has the fourth highest population also has a high urban population but a relatively lower literacy rate. Senegal also has a comprehensive university, but has lower levels of primary and secondary school enrollment than Cameroon and Cote d'Ivoire. During the period of the study, 1995–2002, all six countries participated to

[1] Although Cameroon is geographically situated in Central Africa, it has always been included in Western African countries for scientific studies.

Table 1 Demographic background of the countries

Countries	Population (millions)	% Living in urban Area	Land area (km^2)	Life expectancy	GDP per capita U$	Literacy rate (%)	HDI	HDI rank 2002
Benin	6.7	45.30	110,620	52.96	1,070.00	55.5	0.411	159
Burkina	13.9	20.50	274,200	43.92	1,100.00	26.6	0.330	173
Cameroon	16.4	51.95	465,400	47.99	2,000.00	75.0	0.499	142
Cote d' Ivoire	17.1	45.35	318,000	45.11	1,520.00	59.8	0.396	161
Mali	10.5	29.20	1,240,198	52.35	930.00	46.4	0.337	172
Senegal	10.5	50.28	192,530	52.31	1,580.00	39.2	0.430	156

Source CIA world fact book 2004

differing degrees in programs promoting investments in ICT for African development espoused by UNDP, the World Bank, and other international organizations. Consequently, they can serve as a meaningful sample for comparative analysis of the performance of their ICT infrastructure expansion programs. Table 1 summarizes some demographic data and their performance on the HDI measures.

3 Data Collection

The data on which this analysis is based were gathered from three different sources, the International Telecommunication Union (ITU), the United Nations (UN), and the African Telecommunications Union (ATU). The ITU provides the research community with complete statistical data collected over the years for the telecommunication sector for all countries. The second source, the United Nations provides researchers with reliable data regularly on social and economic indices for a broad set of categories for all countries. The third source is the African Telecommunication Union (ATU). Sibling sister of the ITU, ATU, has a database of a data specifically concerning the African countries. The data of this study were drawn from these sources and cover the period of 1995–2002. This eight-year period has seen the highest investments in ICT on the African continent. The input and output variables used in the DEA analysis of the ICT sectors of the six countries are given in Table 2.

Table 2 List of variables used in DEA analysis

Input variables	Output variables
Population	Revenue from ICT
Number of households	Number of internet users
Investment in ICT	Percentage of households with a telephone
Number of ICT staff	Total telephone traffic
	Number of cellular phones
	Number of main telephone lines

Table 3 CRS technical efficiency scores and ranking

Rank	DMU	1995 (%)	1996 (%)	1997 (%)	1998 (%)	1999 (%)	2000 (%)	2001 (%)	2002 (%)	Average score (%)
1	Cote d' Ivoire	**100.00**	95.24	96.34	**100.00**	**100.00**	**100.00**	**100.00**	**100.00**	98.95
2	Cameroon	95.59	95.42	94.65	97.98	97.54	90.90	**100.00**	**100.00**	96.51
3	Mali	83.27	83.03	82.91	87.35	93.00	96.95	99.69	**100.00**	90.78
4	Senegal	89.76	87.82	**100.00**	91.23	91.38	95.93	**100.00**	**100.00**	94.52
5	Burkina	88.67	83.27	81.68	90.19	**100.00**	89.40	**100.00**	96.01	91.15
6	Benin	82.90	**100.00**	86.83	83.10	84.14	87.05	92.80	**100.00**	89.60

4 Findings and Interpretations

As stated above, we use the CRS input-oriented model to determine the rank of the countries from best practice to worst practice with regard to utilization of ICT investments for infrastructure expansion. As we observe in Table 3, Cote d'Ivoire is by far the most efficient country in the group. It can be considered the 'best practice' country, with an average efficiency score of 98.95 % for the entire period of the study. It is important to note that Cote d'Ivoire has ranked first 6 of the 8 years of the study. It has attained 100 % efficiency 4 times in the years 1998, 2000, 2001, and 2002 and has been the benchmark 38 times in the model. Cameroon ranks second with an overall average of 96.51 %. However, Cameroon operated at 100 % efficiency only twice during the 8 years of the study, in 2001 and 2002. Senegal ranks third in the CRS analysis with an average technical efficiency score of 94.52 % for the period of the study. Senegal operated at 100 % efficiency in 1997, 2001, and 2002 but was never cited as a benchmark. Burkina, Mali, and Benin rank, respectively, fourth, fifth, and sixth with and average score of 91.15 % for Burkina, 90.78 % for Mali, and 80.60 % for Benin. Of these three countries, Benin operated at 100 % efficiency in 1996, 2001, and 2002 and Burkina in 1999, 2001, and 2002. However, in 2001 and 2002, all of the countries operated at 100 % efficiency. Except for Cote d'Ivoire, all of the countries could have been more efficient in the use of their ICT investments. More in-depth analysis would be needed to determine how exactly these countries could have changed their strategies to optimize ICT expansion.

4.1 ICT Revenue Performance Analysis

As a general observation, all six countries show improvements in ICT revenue relative to their ICT investments over the period of time (1995–2002) of this study. We can separate the countries into two groups based on revenue performance, high performance and moderate performance. The high performance group is composed of the three countries, namely Cote d'Ivoire, Senegal, and Cameroon. These countries have invested considerable amounts of capital and ICT labor. Of these three, Senegal has the highest investment rate relative to its GDP. The moderate performance group of countries is composed of Burkina Faso, Mali, and Benin, had relatively low investments in both ICT capital and ICT personnel. The second observation that stands out from this analysis is that of the general trend of the investment curves is rising. They rise constantly from 1996, until 1999 when they slow down slightly, and then continue to rise with some ups and downs. In terms of ICT revenue performance, Senegal has highest rate of revenue relative to GDP. In 1995, Senegal's revenue as a percentage of GDP was 2.4 %, and it rose consistently to attain 4.9 % in 2002 (Fig. 1).

The country with the second highest revenue as a percentage of GDP is Benin. Although Benin has a relatively weak investment in ICT, the revenue from ICT rose consistently between 1996 and 2002. Benin's ICT investments range from 0.8 % of the GDP in 1996 to 1.1 % in 2002. Likewise Benin's ICT revenue rose from 1.6 to 2.8 % over the same period. Although Cameroon is the second largest investor in ICT after Cote d'Ivoire in terms of capital (constant dollars) and ICT

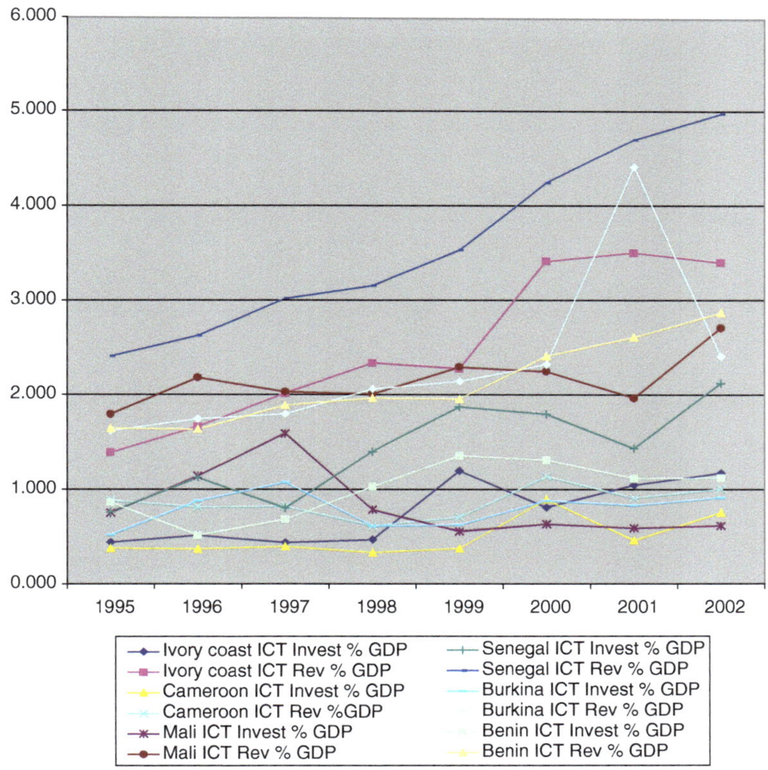

Fig. 1 ICT revenue performance to ICT investments as a percentage of GDP

staff (see Table 3), it earned surprisingly low revenue from its ICT investments compared to the others. Cameroon has the lowest rate of ICT revenue relative to ICT investments for the period 1995–2002. Cote d'Ivoire has a fairly good ratio of returns on ICT investments compared to its investments (in dollars); however, one might expect better returns given the level of its investments. It is not clear why Senegal has a higher return on ICT investments than Cote d'Ivoire. Further investigation in the way investments are used and the very structure of the overall economy of the two countries will help us understand this.

4.2 ICT Infrastructure Expansion

With regard to ICT infrastructure expansion, the performance of the countries varies more widely. Three of them, Cote d'Ivoire, Benin, and Cameroon have been more successful than the others (Fig. 2). However, Cote d'Ivoire stands out for its explosive expansion in the number of cellular telephones and considerable expansion in the area of landline capacity. Starting from 0 cellular telephones in

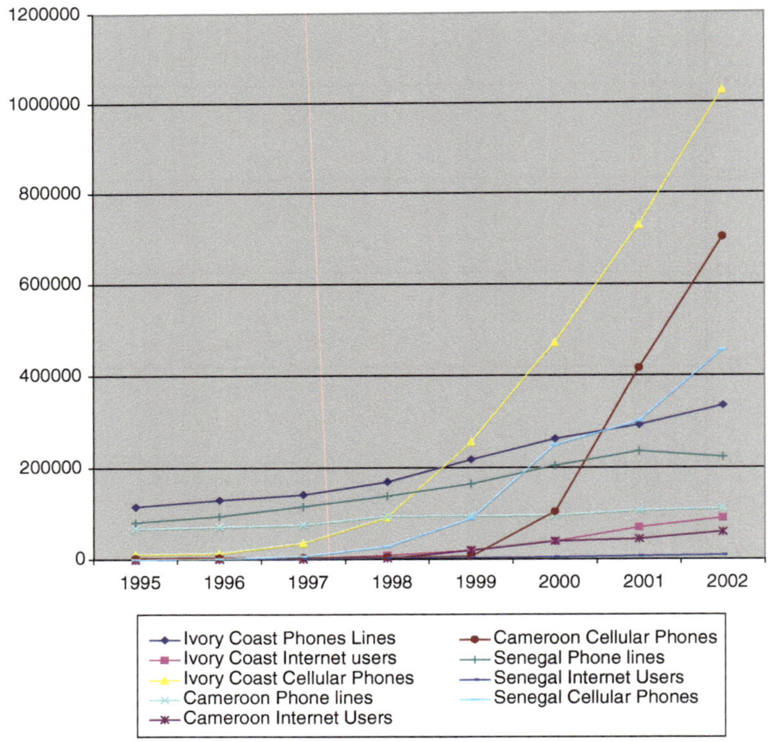

Fig. 2 ICT infrastructure expansion in Cote d'Ivoire, Cameroon, and Senegal

1995, this segment grew gradually to just under 100,000 in 1998 then skyrocketed to just over 1 million in 2002, making Cote d'Ivoire the largest cellular telephone market of the six countries. Cote d'Ivoire also led progress in the area of telephone landlines, increasing from just above 100,000 in 1995 to above 350,000 in 2002. Since 1993, the number of telephone mainlines has been increasing at an average rate of 1.2 % each year and outgoing telephone traffic has been almost doubling every 3 years. Cameroon, the country with the second most rapid expansion in cellular telephones, had very low numbers of cellular telephones between 1995 and 1999. In 1999, this sub-sector started a growth spurt that resulted in an increase to about 100,000 in 2000, then a more rapid rise to around 700,000 cellular telephones.

Senegal had the second most expansion of telephone landlines, behind Cote d'Ivoire, but the third largest expansion in cellular telephones. In 1995, Senegal had just under 100,000 landlines but increased slowly doubling its capacity by 2000 when it leveled off. In the area of cellular telephones, Senegal started from a negligible number of cellular in 1995 to around 240,000 in 2000 then climbed higher to around 450,000 in 2002. Expansion of Internet users has been significantly slower for all three of these countries. All of them report less than 100,000 Internet users in 2002.

The countries, Benin, Burkina Faso, and Mali realized only modest expansion in their ICT infrastructure for the period of the study (Fig. 3). In the area of cellular telephones, Benin is the leader in this group. From 1995 to 1999, it reported

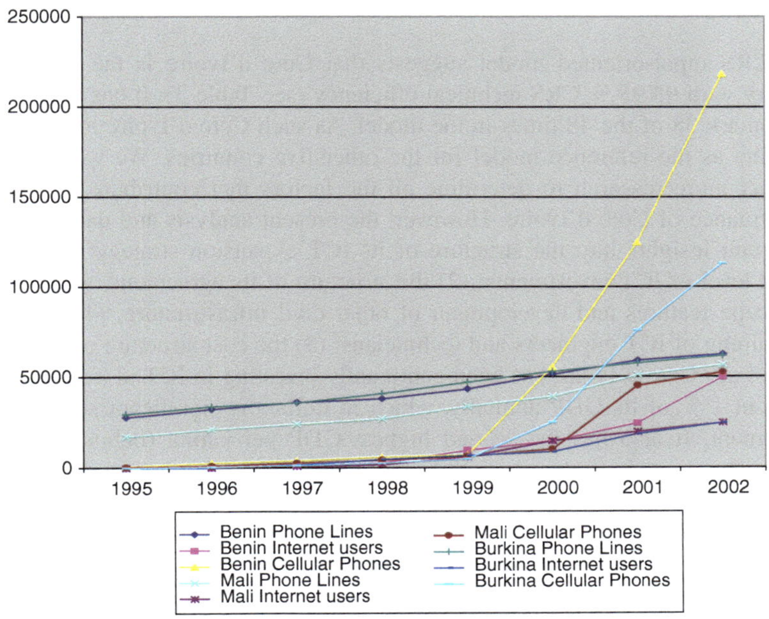

Fig. 3 ICT infrastructure expansion in Benin, Mali, and Burkina

less than 5,000 cellular telephones; but expansion took off in 1999 resulting in a rapid rise to about 225,000 cellular telephones in 2002. It is important to note here that Benin has about one-third of the population and about two-thirds of the GDP per capita of Cote d'Ivoire. The second most cellular telephone expansion for this group took place in Burkina Faso. This country has just about 2.2 times the population of Benin and about 3 times the GDP. Between 1995 and 1999, cellular telephone expansion in Burkina Faso was slow reaching only a few thousand, but between 1999 and 2002 it rose to about 110,000. Mali had the third most increase in cellular telephones moving from just below 5,000 in 1999 to just above 50,000 in 2002. Mali does have a higher population than Benin, but the lowest GDP per capita of the countries in the study. In all three countries, there has been limited expansion of telephone land through the period of this study. Benin almost doubled its telephone landline capacity from about 30,000 in 1995 to just under 60,000 in 2002. Burkina Faso had an almost identical performance over the same period, moving from about 31,000 landlines in 1995 to about 60,000 in 2002. Mali had a more rapid growth in telephone landlines than Benin and Burkina Faso, but started lower with just under 20,000 in 1995 but jumped to about 55,000 in 2002. All three of these countries had low expansion of internet users, while Benin hit the 50,000 mark in 2002, the others achieved only around 25,000 internet users by 2002.

4.3 The Benchmark and Best Practice Country

The CRS input-oriented model suggests that Cote d'Ivoire is the best practice country with 98.95 % CRS technical efficiency (see Table 3); it has also been the benchmark 38 of the 48 times in the model. As such Cote d'Ivoire merits a closer examine as the reference model for the other five countries. We would need do conduct more research to determine all the factors that contribute the excellent performance of Cote d'Ivoire. However, the present analysis and data offer some important insights into the structure of its ICT expansion strategy: (1) The sustained level of ICT investments; (2) the structure of its agreements with FCR; (3) landscape features and development of other civil infrastructure; (4) its capacity for training of ICT engineers and technicians; (5) the cost structure of its services. First of all, Cote d'Ivoire has been continually investing in ICT at a sustained rate of about 1 % of its GDP annually, which in dollars is significant sum of capital investment. It also has the second highest GDP per capita (behind Cameroon) among the countries being studied. Cameroon and the other countries which have lower GDPs have been investing proportionally smaller amounts of their GDP in ICT. The exception to this trend is Senegal, which has been investing more than Cote d'Ivoire as a proportion of GDP. However, since Cote d'Ivoire has a much higher GDP than Senegal, Senegal's investment in dollar terms is much less than that of Cote d'Ivoire. For the period of the study, Cote d'Ivoire had the second highest ICT revenue per dollar of investment behind Senegal. From the CRS analysis, we observe that it has the best good utilization of ICT investments and returns. All the other countries have exhibited levels of inefficiency in the utilization of their ICT investments.

Second, relative to the other five countries, Cote d'Ivoire had developed a unique ICT expansion strategy. In 1997, CI-Telecom was fully privatized, with the state retaining 35 % of its shares while 51 % were bought by France Cable and Radio (FCR). As part of the deal, FRC committed to invest USD 417 million to the expansion of the telecommunications infrastructure over five years which led to the addition of 290,000 landlines to the existing 140,000. The other countries made no such deals when they privatized their telephone sectors. Consequently, ICT capital reinvestment has been a low priority in these countries, limited to network maintenance and incremental growth. A third factor that favors Cote d'Ivoire's landline expansion is its excellent civil infrastructure, relatively small land area (318,000 km^2) and high density of urban population (45.4 %). Cote d'Ivoire has 68,000 km of roads of which 6,500 is paved. Cameroon is a bit larger in land area (465,400 km^2) than Cote d'Ivoire and has the highest urban population (51.96 %). However, it ranks behind Cote d'Ivoire in developed civil infrastructure, having 50,000 km of roads of which only 4,300 are paved. In contrast to these two, countries such as Senegal, Mali, and Burkina Faso have relatively lower levels of civil infrastructure development and smaller urban populations which are both impediments to the expansion of ICT landlines. Although Senegal has the second highest urban population (50.28 %) and the second smallest land area (192,530 km^2), it has only 2,678 km of roads of which 395 km are paved. Burkina Faso has a little larger land area (274,000 km^2) than Senegal but the lowest urban population (20.5) of all the countries in the study. Its civil infrastructure comprises 15,272 km of roads of which 2,416 is paved. Mali on the other hand has a vastness and inhospitable landscape that is a formidable impediment to the development of a large landline network. Mali comprises some 1,240,198 km^2 of mostly desert and has only 29.25 of its population living in urban areas. Benin is the smallest country of the six with a total land area of 110,620 km^2, the fourth largest urban population (45.3 %), a total of 3,500 km of roads of which 1,195 km is paved. It is clear that the level of civil infrastructure development was a factor that aided ICT expansion in Cote d'Ivoire. Further, the electricity distribution sector of Cote d'Ivoire is now developing and testing approaches to ICT signal transmission over their infrastructure.

The fourth important factor in Cote d'Ivoire's ICT expansion performance is its capacity to educate annually some 30,000 ICT engineers and technicians at its ten tertiary institutions for technological education.[2] This capacity to produce a large number of high trained ICT professionals is unparalleled in the other five countries.

---

[2] For example, Ecole Nationale Superieure d'Ingenieurs, Ecole de Technologie Tertiaire, Ecole Nationale des Techniciens Superieurs, the Centre de Formation Continue, Ecole Superieure Interafricaine de l'Electricite, Institut National Polytechnique Houphouet-Boigny, Ecole National Superieure des Travaux Publics and Ecole Preparatoire.

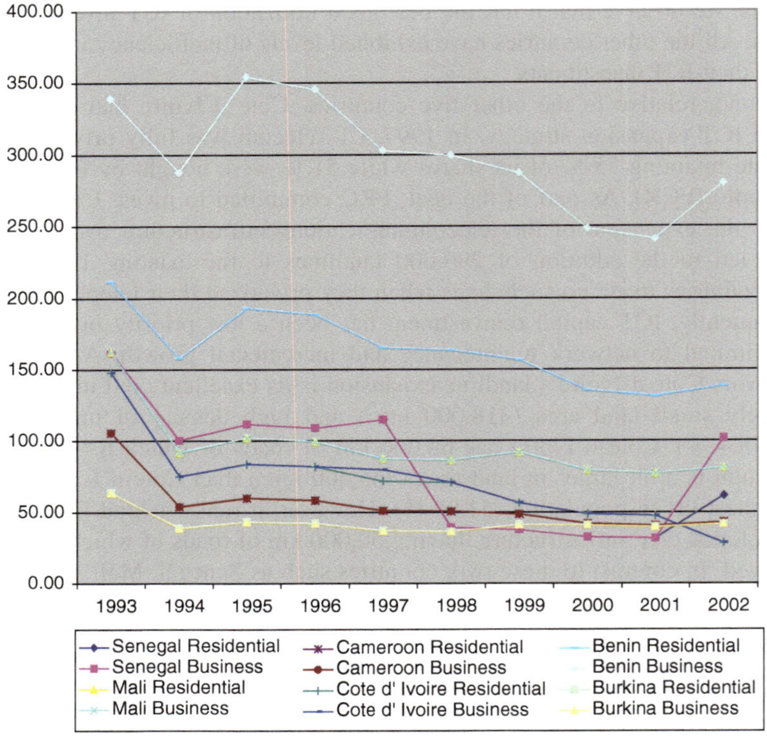

Fig. 4 Hook-up cost for business and residential telephone lines

On the contrary, Cameroon and Senegal each have only one polytechnic institution with limited capacity for training ICT engineers and technicians, and the others have none. Essentially, most of the aspiring ICT engineers and technicians seek training in Cote d'Ivoire where most of them remain and work after completing their studies. Perhaps this is why its ICT sector has ten times the number of specialist employed that Benin, twice that of Cameroon, and three times that of Senegal. The ability of Cote d'Ivoire to train its own staff gives it an advantage over the other five countries. It is no surprise that Cote d'Ivoire lead these countries in the low level of failures per 100 telephone circuits. Further as it has been reported elsewhere complementary investments in education, healthcare, and other civil infrastructures lead to higher impact of ICT on human development.

A fifth important factor that differentiates Cote d'Ivoire from the other countries is its service-cost structure. All the countries in the study are economically challenged; their GDP per capita during the period of the study did not rise above U$2500 (see Appendix A). Consequently, any change in the service-cost structure is likely to cause dramatic shifts in consumer response. For example, in Cote d'Ivoire the cost to acquire business or residential telephone line has dropped from U$150 in 1995 to less than U$50 in 2002 (Fig. 4), while in the other countries it has stayed the same or rose during the same period. In the case of Cameroon, these costs have remained unchanged for the entire period; in Senegal, it went up by 50 % between 1995 and 1999 then dropped back down by about 60 % then rose again to just below the 1995 levels. In Benin and Burkina Faso, these charges rose 50 % from 1995 to 2002, and for Mali they went up 20 % over the period. This decline in telephone cost in hoop-up cost in Cote d'Ivoire has undoubtedly led to increasing demand for telephone landline service. A second aspect of the service-cost structure that has had a dramatic effect in ICT in Cote d'Ivoire is the cost of initiating cellular telephone service. From 1998 to 1999, the cost initiating cellular telephone service dropped by 24 % then again by another 31 % between 1999 and 2000. Following the reductions in 1998, cellular telephones in Cote d'Ivoire increased from 91,000 subscribers to 257,000 subscribers in 1999. Then in conjunction with the drop in cost from 1999 to 2000 cellular telephones increased again to reach 472,952 units, they then continued to increase reaching 1 million in 2002.

It is clear that the price structure of ICT services over the period of the study assisted the rapid expansion in Cote d'Ivoire. On the other, Cameroon which had the second highest growth in cellular telephone service also had the highest per capita GDP for the entire period of the study. During 1998 and 2001, Cameroon reduced its cellular service connection price by 90 %; during this same period, it realized an increase in cellular telephones subscribers from 5,000 to 417,295. Benin, however, had a contrary experience; from 1999 to 2002, its cellular service connection charges increased by 500 % and its monthly service charges also increased by 100 %, but it still managed to achieve expansion in the number of cellular telephone subscribers from 7,269 to 218,770 in the same period. In Burkina Faso, cellular telephone connection charges gradually declined by 33 % from 1998 to 1999, then by a further 23 % in 2000, and 50 % in 2002. Along with these price changes, the number of cellular telephone subscribers in Burkina rose from 2,730 in 1998 to 5,036 in 1999, to 25,245, 76,000, and 113,000 in 2000, 2001, and 2002, respectively. In 1995, Senegal had a total of 122 cellular telephone subscribers, and then a 30 % reduction in the connection cost resulted in an increase to 1,412 telephone subscribers in 1996. In 1997, there was a further 50 % reduction followed by a rapid increase in cellular telephone subscribers, to 6,942 in 1997, then 27,487 in 1998 then a steep rise to 301,000 by 2002.

5 Conclusions

This research provides some interesting empirical evidence concerning the impact of ICT investments on the expansion of the ICT sectors of these African countries. The choice of DEA for this analysis was deliberate because I did not have enough data to conduct reliable regression analysis to determine the exact correlations between the different factors and the expansion in ICT infrastructure. However, with the DEA method, I was able to clearly answer the two basic questions of this research: (1) Are investments in ICT technical efficient with regard to the building and expansion of ICT infrastructure? (2) Are investments in ICT resulting in revenue growth and contributing to growth in GDP (a component of development)? For Cote d'Ivoire the evidence is clear on both questions, investments in their ICT are clearly technically efficient with regard to ICT infrastructure expansion and are contributing to GDP growth. This analysis also demonstrated that Cote d'Ivoire is a benchmark country that the other five could emulate to improve their performance. Cameroon and Senegal also showed some noteworthy gains in ICT expansion from their investments; however, more in-depth investigations into the ICT strategies and social conditions of these countries could yield important insights that would help to reformulate their ICT investment strategy and policies with a view to improving the technical efficiency of their ICT sectors. Further, Benin, Mali, and Burkina are not efficiently utilizing their investments in ICT to achieve any significant returns in terms of infrastructure expansion and consequent social development. Some of the reason for not achieving higher technical efficiency might have to do with their cost structure of their ICT services and/or limited civil infrastructure of some of these countries. In the case of Mali, its vast inhospitable landscape, low GDP per capita, and low urban population are likely to be continued obstacles to ICT expansion. Presently, we are not in a position to verify that these factors are significant obstacles to ICT expansion. More research is needed to verify the validity of this hypothesis. However, this analysis has pointed to five factors: (1) The sustained level of ICT investments; (2) the structure of its agreements with FCR; (3) landscape features and development of other civil infrastructure; (4) its capacity for training of ICT engineers and technicians; (5) the cost structure of its services that distinguish Cote d'Ivoire as the best practice country. And although I cannot provide the direct proportional impact of these five factors on the ICT expansion performance of Cote d'Ivoire, it is clear that these need to be considered by the other countries as the reformulate new ICT policies. Consequently, it would be informative for the other five countries to examine more carefully the structure of ICT strategy in Cote d'Ivoire with a view to improving their own. The results of this study can also be taken one step further with other methods such as regression analysis. This could help bring another dimension of confirmation of these results. Finally, this research makes a contribution to African ICT policy makers by providing both evidence and an approach to analysis that could inform their work as the embark on the formulation of new development strategies. This DEA and quantitative approach is a powerful methodology that could help policy makers conduct interesting and comprehensive analysis for decision making.

Acknowledgment Material in this chapter previously appeared in: *Electronic Journal of Information Systems in Developing Countries* 26:5, pp. 1–16.

Appendix A

Countries	1995	1996	1997	1998	1999	2000	2001	2002
Benin	800.00	840.00	880.00	900.00	930.00	980.00	1030.00	1070.00
Burkina Faso	850.00	900.00	950.00	950.00	1,000.00	1,020.00	1,060.00	1,100.00
Cameroon	1,540.00	1,600.00	1,680.00	1,720.00	1,770.00	1,880.00	1,960.00	2,000.00
Cote d'Ivoire	1,440.00	1,510.00	1,560.00	1,610.00	1,630.00	1,590.00	1,590.00	1,520.00
Mali	640.00	660.00	700.00	740.00	780.00	740.00	890.00	930.00
Senegal	1,230.00	1,260.00	1,310.00	1,360.00	1,420.00	1,490.00	1,570.00	1,580.00

GDP per capita, current international dollar, *Source* UN Statistics Division Database

Appendix B

Countries	1993	1994	1995	1996	1997	1998	1999	2000	2001	2002
Senegal residential	162.15	100.86	112.19	109.47	115.13	39.66	38.01	32.87	31.92	61.71
Senegal business	162.15	100.86	112.19	109.47	115.13	39.66	38.01	32.87	31.92	101.88
Mali rèsidential	162.45	91.86	102.17	99.70	87.38	86.45	91.81	79.39	77.11	81.10
Mali business	162.45	91.86	102.17	99.70	87.38	86.45	91.81	79.39	77.11	81.10
Cameroon residential	105.95	54.03	60.10	58.65	51.40	50.85	48.73	42.14	40.93	43.04
Cameroon business	105.95	54.03	60.10	58.65	51.40	50.85	48.73	42.14	40.93	43.04
Cote d'Ivoire residential	148.33	75.65	84.14	82.10	71.96	71.19	56.85	49.16	47.75	28.69
Cote d'Ivoire business	148.33	75.65	84.14	82.10	79.95	71.19	56.85	49.16	47.75	28.69
Benin residential	211.89	157.99	193.31	188.63	165.32	163.56	156.72	135.53	131.64	138.45
Benin business	339.03	287.67	353.99	345.41	302.73	299.50	286.98	248.17	241.04	279.77
Burkina residential	63.57	38.90	43.27	42.22	37.01	36.61	41.40	41.17	39.99	42.05
Burkina business	63.57	38.90	43.27	42.22	37.01	36.61	41.40	41.17	39.99	42.05

Cost of Business and Residential Telephone line connection charges, *Source* ITU Yearbook

References

Akpan P (2000) 'Africa and the new ICTs: implications for development'. Paper presented at the International Federation for information processing conference, Cape Town, South Africa, May 2000

Avgerou C (1998) How can IT enable economic growth in developing countries? Inf Technol Dev (8):15–29

Braga CA et al (2000) The networking revolution: opportunities and challenges for developing countries (working paper). World Bank Group, Washington DC. Retrieved Oct 2000 from http://www.infodev.org/library/working.html

Chen Y, Zhu J (2004) Measuring Information technology's indirect impact on firm performance. Inf Technol Manage J (5):9–22

Färe R, Grosskopf S, Lovell C (1983) The structure of technical efficiency. Scand J Econ (85):181–190

Landauer T (1995) The trouble with computers, usefulness, usability, and productivity. MIT Press, Cambridge

Mwesige PG (2004) Cyber elites: a survey of internet café users in Uganda. Telemat Inf 21(2004):83–101

Shafer S, Byrd T (2000) A framework for measuring the efficiency of organizational investments in information technology using envelopment analysis. Omega: Int J Manage Sci (28):125–141

UNDP (2001) Making new technologies work for human development

Wang C, Gopal R, Zionts S (1997) Use of data envelopment analysis in assessing information technology impact on firm performance. Ann Oper Res (73):191–213

Chapter 13
A Hybrid DEA/DM-Based DSS for Productivity-Driven Environments

Sergey Samoilenko and Kweku-Muata Osei-Bryson

This chapter begins with the overview of the challenges facing organizations competing in dynamic business environments and the associated demands placed on organizational DSS. The overview of the design of DSS starts with an outline of the capabilities and combination of such various methods as data envelopment analysis (DEA), cluster analysis (CA), decision tree (DT), neural networks (NN), and multivariate regression (MR) offer to a decision maker. In a step-by-step fashion, the reader is introduced to the various modules of the system that are eventually combined into a comprehensive DSS. An extensive illustrative example offers the reader an opportunity to observe the DSS in action.

1 Introduction

Modern organizations typically operate in dynamic, competitive environments. Within this context, the critical issues of organizational survival and advancement often lead to calls for improvements in the levels of effectiveness and efficiency of performance. However, due to the relativity of the concepts of efficiency and effectiveness, productivity-driven organizations must take into consideration the performance of their competitors. This requirement is due to the dynamic nature of the business environment, which will cause the levels of performance of

S. Samoilenko (✉)
Averett University, Department of Computer Science, 420 W. Main Street, Danville, VA 24541, USA
e-mail: SSamoilenko@Averett.Edu

K.-M. Osei-Bryson
Virginia Commonwealth University, Department of Information Systems, 301 W. Main Street, Richmond, VA 23284, USA
e-mail: KMOsei@VCU.Edu

K.-M. Osei-Bryson and O. Ngwenyama (eds.), *Advances in Research Methods for Information Systems Research*, Integrated Series in Information Systems 34, DOI: 10.1007/978-1-4614-9463-8_13,

competing organizations to change over time, and if the efficiency of the competitors has improved, then a productivity-driven organization must respond with its own improvements in efficiency.

A desired capability of an organization to successfully respond to efficiency-related challenges suggests the need, first, for an effective mechanism that allows for discovering appropriate productivity models for improving overall organizational performance and, second, for a feedback-type mechanism that allows for evaluating multiple productivity models in order to select the most suitable one.

The dynamic nature of the business environment also suggests the presence of a concept that is central to a productivity-driven organization, namely that of the *superior stable configuration*. Given the goal of achieving a high level of efficiency of conversion of inputs into outputs, a superior stable configuration in the context of a productivity-driven organization may imply *a model of conversion of inputs into output (input–output model) characterized by a high level of efficiency*.

Overall, a decision maker tasked with a responsibility of improving performance of productivity-driven organization existing within a dynamic business environment must take into consideration internal (organizational) and external (environmental) factors. Similarly, if an information system is to adequately support such a decision making context then the designers of such a system must implement two sets of functionalities: *externally oriented* and *internally oriented*. The *externally oriented* functionality is directed toward evaluating the external competitive environment of a productivity-driven organization, as well as identifying the differences between the current state of the organization and the states of its competitors. The *internally oriented* functionality, on the other hand, is directed toward the optimization of the level of productivity of the organization, as well as toward an identification of the factors impacting the efficiency of the input–output process.

In this chapter, we will describe a decision support system (DSS) that allows assessing and managing the relative performance of organizations. Specifically, we focus on organizations that consider the states of their internal and external organizational environment in the formulation of their strategies, such that the achievement of an organizational goal is dependent on the level of performance that is commonly measured in terms of the levels of the efficiency of utilization of inputs, effectiveness of the production of outputs, and efficiency of conversion of inputs into outputs.

2 Description of the DSS

The focus on the efficiency assessment suggests that an important component technique of the DSS is DEA. However, other techniques are also required for providing answers to several questions that are relevant to the organization's search for the productivity model that is most suitable with respect to survival and advancement. Next, we outline how a DSS could be implemented using a combination of parametric and nonparametric data analytic and data mining techniques including DEA, CA, DT, NN, and MR.

Let us discuss how each of the above-mentioned methods, alone, or in combination with other methods, could be used in the DSS. First, we will discuss externally oriented functionality.

2.1 Externally Oriented Functionality

Cluster analysis allows for segmentation of the data set into naturally occurring heterogeneous groups. An application of this method allows for detecting the presence of multiple disparate groups of competitors in the external business environment. A decision maker can also determine whether the clusters comprising the data set differ in terms of the relative efficiency of utilization of inputs or production of outputs—DEA will help in this regard. By specifying a DEA model and running the analysis, we can obtain scores of the relative efficiency for each cluster, as well as see how the score differs between the clusters.

If the data for the same group of competitors available for two points in time, let us say, *Year 1* and *Year 2*, then a decision maker can obtain insights regarding possible changes in the number of clusters, as well as changes in the membership of the clusters. This will allow for determining whether the structure of the competitive environment has changed over time. If DEA follows CA, then a decision maker can also identify the changes that took place in regard to relative efficiencies of the members comprising the data set.

After conducting CA and DEA, a decision maker may want to inquire into the existence of the factors that are possibly responsible for the presence of multiple groups of competitors, namely what are some of the reasons for the heterogeneity of the external business environment? DTs can offer some help in this regard; once CA helped to identify the presence of the various groups comprising the sample, a target variable (let us say, *Cluster#*) will allow for identifying every member of the data set in terms of its membership in a given group. Then, we can run DT analysis specifying that target variable—this will allow for determining the dimensions in the data set that differentiate the clusters the most. By comparing the results of DT analysis of the data set representing two points in time, we can determine, based on changes in differentiating dimensions, the possible reasons for the changes not only in the number of clusters, but also in the composition of the clusters.

By creating a new target variable reflecting the differences in the scores of the relative efficiency of each cluster (e.g., *LessEfficient* vs. *MoreEfficient*), we can also use the same DT-based approach to identify the differences between the clusters in regard to the scores of the relative efficiency. By conducting the same analysis at multiple points in time, we can identify the changes in regard to the relative efficiency, as well as factors possibly associated with identified changes.

At this point, we have obtained insights regarding the nature of the competitive environment—we identified the naturally occurring groupings, as well as changes in the groupings over time. We determined some of the variables that are responsible for separating the clusters, as well as the variables that were responsible for change in groupings over time. We also found out important information regarding the disparity between the clusters in terms of the relative efficiency, as well as some of the variables that are possibly responsible for the disparity.

Let us summarize the capabilities afforded by various combinations of CA, DEA, and DT to a decision maker for the purposes of assessing the external environment of an organization.

When the data analysis is conducted at *one point in time*:

- A combination of CA and DT allows for identifying naturally occurring groups of competitors, as well as determining major dimensions that differentiate the groups.
- A combination of CA and DEA allows for comparison of the naturally occurring groups in terms of the relative efficiency of conversion of inputs into outputs.
- A combination of CA, DEA, and DT allows for identifying factors differentiating relatively less efficient and relatively more efficient groups of competitors.

Additionally, when the data analysis is conducted at *multiple points in time*:

- A combination of CA and DT allows for identifying changes that took place in regard to the composition of naturally occurring groups of competitors, as well as for determining the factors associated with the changes.
- A combination of CA and DEA allows for identifying the changes that took place in regard to the relative efficiency of the naturally occurring groups.
- A combination of CA, DEA, and DT allows for identifying factors associated with the changes in relative efficiency of the group that took place over time.

2.2 Internally Oriented Functionality

Once we completed the analysis of the external environment of the organization, we need to turn our attention to the analysis of its internal state. All organizations are similar in the way that every one of them converts resources into products or services—this allows a decision maker to model an organization as a set of inputs (e.g., resources) and a set of outputs (e.g., products and/or services). MR analysis allows us to identify those independent variables (inputs) and their interaction terms (complementarities) that produce a statistically significant impact on the dependent variable (output). It is, of course, a limitation of MR that the analysis allows for a single output variable. However, if a decision maker wants to consider multiple outputs, then this requirement could be accomplished using multiple MR models.

Additionally, we can use NN to create a model of input–output transformation of the organization. This requires identifying a set of inputs and outputs—the two sets will become, respectively, input and output nodes of NN. Once NN is trained, we can identify an *input–output transformation* model specific to the organization.

By saving this model and varying levels of inputs, we can determine the impact of the manipulation of inputs on the level of outputs. More importantly, we can use NN for the purposes of benchmarking—this will require creating a NN model of a better performing peer. In doing so, we can apply existing levels of the inputs of the organization to the transformation model of the better performing organizations to discern whether the improvements in performance could be obtained via the variation of the levels of inputs, or via improvements in the input–output transformation model.

Moreover, once we obtained via NN a set of simulated inputs and outputs, we can subject the set to DEA. In this situation, a decision maker has two options in regard to input and output variables. First, the same set of inputs and outputs model could be used for a DEA model and NN analysis. Second, the variables used in DEA model could represent a subset of inputs and outputs used in NN analysis. In either case, the simulated set of inputs and outputs could be used to create a representation of two *simulated organizations*. The first simulation represents a level of outputs based on the manipulated inputs, but with the original input–output transformation process. The second set represents a set of outputs based on the original set of inputs and manipulated input–output transformation process that was adapted from a better performing organization. Adding the simulated data to the original data set and then running DEA will allow for determining the impact of the simulation on the relative efficiency of utilization of inputs and production of outputs. As a result, a decision maker will be able to determine whether the improvements in efficiency should be obtained via variation in the levels of inputs, or whether the improvements should come from the changes in input–output transformation process.

At this point, we can summarize the capabilities afforded by utilization of MR, DEA, and NN to a decision maker in regard to the analysis of the internal state of an organization.

When the data analysis is conducted at *one point in time*:

- MR allows for determining resources that have a significant impact on the output of the business process, as well as for determining the presence of synergies and complementarities of the resources on the output.
- NN allows for determining whether the level of the output of the business process should be increased by means of increasing the level of inputs, or by means of improving the process of converting resources into outputs.
- Combination of NN and DEA allows for determining whether the improvements in the relative efficiency of the organization should come from changes in the level of consumed resources, or from the changes in the business process by which resources are converted into outputs—products or services.

Also, when the data analysis is conducted at *multiple points in time*:

- MR provides insights into the possible changes in regard to significance of the impact of resources, as well as their complementarities, on the outputs of the business processes.
- NN offers indications whether the process of input–output transformation of the organization has changed over a period of time.

- DEA and NN provide evidence regarding the changes that took place in regard to the relative efficiency of the organization, as well as offer insights into whether the identified changes are due to the changes of the levels of inputs and outputs, or due the changes in the organizational input–output transformation process.

2.3 Architecture of the DSS

At this point, we can combine the sequences of the data analytic methods that were outlined above within a comprehensive design of a single DSS (see Fig. 1).

3 An Illustrative Application

The outlined DSS could be used in the context of any set of economic units, as long as the units are represented via the common DEA model. However, the analytical engine of the DSS should not be limited to a data set that is comprised only of inputs and outputs of the DEA model—a decision maker can obtain a richer set of insights if other variables are also included in the data set. As a consequence of the reliance of the DSS on the common DEA model, the system could be applied at the different levels of granularity—the DSS could be used at the departmental, organizational, industry, of a country level of analysis.

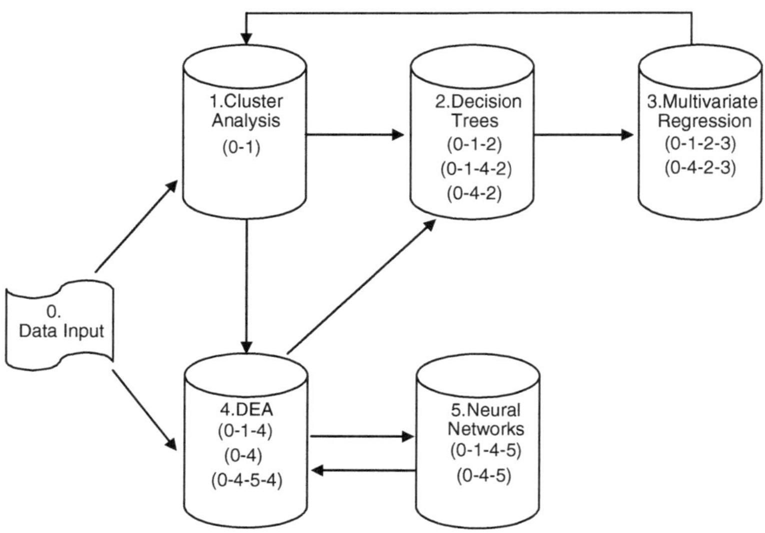

Fig. 1 Sequential method utilization within the design of the hybrid DEA-/DM-based DSS

This illustrative example involves the country level and deals with the efficiency and effectiveness of the impact of investments in telecoms, a type of investment that is common to almost all of the economies in the world. The context is represented by the following 18 economies: Albania, Armenia, Azerbaijan, Belarus, Bulgaria, Czech Republic, Estonia, Hungary, Kazakhstan, Kyrgyz Republic, Latvia, Lithuania, Moldova, Poland, Romania, Slovakia, Slovenia, and Ukraine. The time series data covering the period from 1993 to 2002 were obtained from the *World Development Indicators* database and the International Telecommunication Union' *Yearbook of Statistics.*

Within the context of the sample, 18 economies are economic entities characterized by the same business process—that of conversion of investments in telecoms into revenues. Because economies compete for foreign direct investments and private investments, they are forced to compete with each other based on their levels of productive efficiency of conversion of investments into revenues. Clearly, the more productive economy would attract a larger pool of investment resources than the less productive one. Hence, a decision maker may pose the following general question:

> How could a given economy improve its level of productivity with regards to its investments in telecoms?

Undoubtedly, the posed question is complex, primarily due to the factors impacting the measure of *level of productivity*, *namely* utilization of investments, production of revenues, and transformation of investments into revenues.

Thus, the general question could be expanded into three efficiency-based subquestions:

1. How could a given economy improve its level of efficiency of utilization of investments in telecoms?
2. How could a given economy improve its level of efficiency of production of revenues from telecoms?
3. How could a given economy improve its level of efficiency of the process of conversion of investments into revenues from telecoms?

We use SAS' *Enterprise Miner* data mining software to conduct CA, DT, NN, and MR, and *OnFront* to conduct DEA. Because the design of the proposed DSS systems is DEA-centric, one of the prerequisites for using it is associated with identifying a DEA model that is to be used in evaluating productivity of the organizational entities in the sample. We list the set of variables used in the illustrative example below.

Input variables:

- GDP per capita (in current US $)
- Full-time telecommunication staff (% of total labor force)
- Annual telecom investment per telecom worker
- Annual telecom investment (% of GDP in current US $)
- Annual telecom investment per capita
- Annual telecom investment per worker

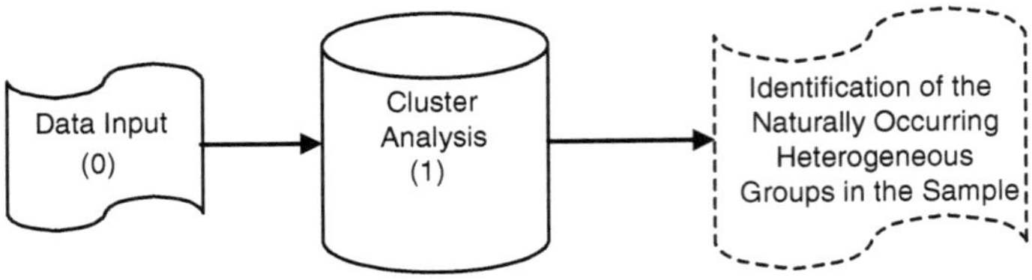

Fig. 2 Detection of changes in the external competitive environment

Output variables:

- Total telecom services revenue per telecom worker
- Total telecom services revenue (% of GDP in current US $)
- Total telecom services revenue per worker
- Total telecom services revenue per capita.

Next, we demonstrate, in a step-by-step fashion, an application of the DSS to the context of our illustrative example.

Step 1 Is the Business Environment Homogeneous?

The purpose of the first step is to offer the decision maker a capability to inquire into the nature of the competitive business environment in regard to the presence of the multiple heterogeneous groups of competing business entities. As the reader may recall, we implement this functionality by incorporating CA into the design of our DSS (Fig. 2).

We began the CA using the "automatic" setting, which did not require a specification of the exact number of clusters by the analyst. This setting produced a five-cluster solution that was considered to be the starting point in the analysis. By sequentially reducing the number of clusters, we derived a two-cluster solution which was considered to be final; the result is provided below.

Cluster1:	Czech Republic, Estonia, Hungary, Latvia, Lithuania, Poland, Slovenia, Slovakia
Cluster2:	Albania, Armenia, Azerbaijan, Belarus, Bulgaria, Kazakhstan, Kyrgyzstan, Moldova, Romania, Ukraine

However, what is the basis for accepting this two cluster segmentation versus some other grouping? We suggest using an external evaluation approach to assess cluster validity, where a domain expert opinion can provide external confirmation of the validity of this segmentation. In the case of the illustrative example, such domain expert support is provided by Piatkowski (2003), who concluded that in the period "between 1995 and 2000 ICT capital has most potently contributed to output growth in the Czech Republic, Hungary, Poland, and Slovenia." Thus, it could be suggested that we were able to separate 18 economies into the two groups: the *Leaders* group (Cluster1) that consists of economies which benefited

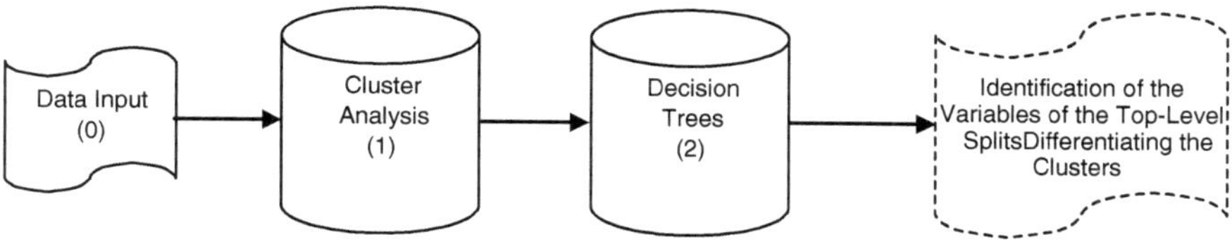

Fig. 3 Identification of the factors that differentiate groups of competitors

the most from the investments in telecom, and the *Followers* group (Cluster2) that consists of economies where the benefits are less pronounced.

Furthermore, the results of the CA offered evidence that Cluster 1 is different from Cluster 2 in terms of the two dimensions: *Investments* and *Revenues from telecoms*. Consequently, with regard to these dimensions, for a given economy its own cluster will represent a peer context, while members of the other cluster will comprise a non-peer context. Overall, Step 1 allowed the decision maker to determine that the competitive business environment comprised of 18 economies is not homogenous, but appears to be comprised of two groups that differ in terms of investments and revenues from telecoms.

Step 2 What are the factors Responsible for Heterogeneity of the Business Environment?

However, even if the decision maker identified the presence of multiple groups comprising a given business environment, it is not clear what differentiates the peer context from the non-peer context; this question will be answered by the functionality of our DSS described in Step 2 (Fig. 3).

The results of the CA allow us to introduce a target variable *Cluster#* to serve as an identifier of a given group in the sample. Using this variable in DT analysis, we can determine, based on the top-level split, the dimension that differentiates the *Leaders* from the *Followers*. Clearly, when conducting DT analysis, we do not have to be limited to the set of variables that was used for CA. The results of DT analysis in the form of the decision rules are presented below.

IF Annual telecom investment (Current US $ per telecom worker) < $9,610 **THEN** Cluster = {2: **96.4%;** 1: 3.6 %} where $N = 110$.
IF Annual telecom investment (Current US $ per telecom worker) ≥ $9,610 **THEN** Cluster = {2: 2.9 %; 1**: 97.1%**} where $N = 70$.

These results allow the decision maker to identify the relevant dimension that differentiates the peer from the non-peer context the most. This means that while the two clusters differ in terms of investments and revenues from telecoms, the single most important dimension that differentiates two clusters is associated with the level of investments in telecoms per telecom worker. Step 2 allows the decision maker to determine that the two groups of 18 economies differ most significantly in terms of the respective levels of investments in telecoms per telecom worker.

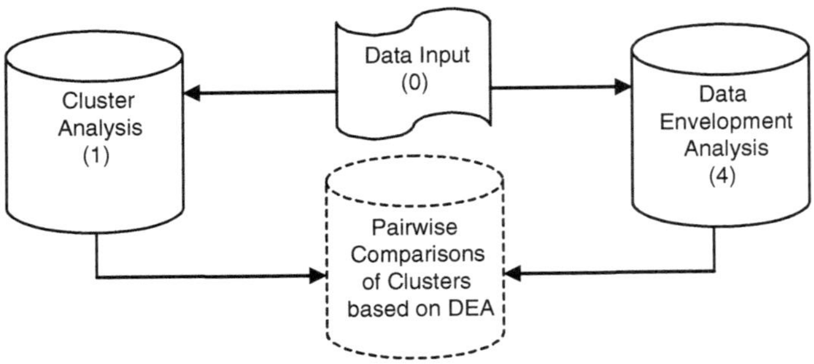

Fig. 4 What are the differences in relative efficiency among peer and non-peer groups?

Table 1 Assessment of the differences between the clusters in terms of the relative efficiency

Orientation	Return to scale	Cluster1	Cluster2	Conclusion
Input oriented	CRS	0.89	0.79	*Cluster1* is relatively more efficient than *Cluster2*
	VRS	0.95	0.88	*Cluster1* is relatively more efficient than *Cluster2*
	NIRS	0.89	0.80	*Cluster1* is relatively more efficient than *Cluster2*
Output oriented	CRS	1.21	1.44	*Cluster1* is relatively more efficient than *Cluster2*
	VRS	1.18	1.30	*Cluster1* is relatively more efficient than *Cluster2*
	NIRS	1.21	1.38	*Cluster1* is relatively more efficient than *Cluster2*

Step 3 Do Groups of Competitors Differ in Terms of the Relative Efficiency?

Once the decision maker identified the presence of heterogeneous groups of the competitors within the business environment, it is reasonable to inquire whether the groups differ in terms of the efficiency of utilization of investment and production of revenues. This additional information is obtained by means of incorporating DEA in the design of the proposed DSS (Fig. 4).

Completing Step 3 involves running DEA and calculating the scores of the relative efficiency for each entity in the sample. It should be noted that DEA is not applied separately to the *Followers* and the *Leaders*. Instead, DEA is applied to the entire sample, and so the relative efficiency scores are not determined based on a cluster membership. Our application of DEA resulted in the scores of the *relative efficiency* for each entity in the entire set. If the scores are averaged per cluster, then the decision maker has the information regarding the averaged relative efficiency for the peer versus non-peer groups. For the purposes of our illustrative example, we conducted DEA under assumptions of *constant* (CRS), *variable* (VRS), and *non-increasing return to scale* (NIRS) and averaged the scores for the *Leaders* (Cluster1) and the *Followers* (Cluster2). The results are presented in Table 1.

During Step 3, we can also conduct DEA to calculate the *Malmquist index* (MI) of productivity growth for both clusters in order to measure changes in the productivity and efficiency over time. This will require evaluating the relative magnitude of the components of MI, namely *change in efficiency* (EC) and *change in technology* (TC). The comparison allows the decision maker to identify whether the growth in productivity was primarily efficiency or technology driven.

Overall, within the context of our illustrative example, Step 3 offers the following information to a decision maker:

- The group of the Leaders is relatively more efficient than the group of the Followers in terms of the utilization of investments and production of revenues from Telecoms.
- Individual members of the group of the Leaders are, on average, relatively more efficient than the Individual members of the group of Followers.
- Both groups contain relatively efficient and relatively inefficient economies.
- The changes in the level of productivity of the Leaders are driven by changes in efficiency, while the changes in the level of productivity of the Followers are driven by changes in technology.

As the reader can see, the results of Step 3 provide the decision maker with important information regarding the relative efficiency of the peer versus the non-peer group within the competitive business environment. In the case of our illustrative example, an investigator can easily determine that under any assumption of return to scale the *Leaders* are relatively more efficient than the *Followers* in terms of, both, utilization of investments and the production of revenues. However, we can expect that each cluster will contain relatively efficient economies and relatively inefficient ones. The purpose of the next step is to provide the decision maker with the functionality allowing inquiry into the differences between relatively inefficient and relatively efficient peer and non-peer economies.

Step 4 What are some of the Factors Associated with the Differences in Relative Efficiency?

Because in the case of our illustrative example, we ended up with two clusters (the *Leaders* and the *Followers*), we can identify four groups of economies within our sample. The groups are as follows: relatively efficient economies of the *Leaders*, relatively inefficient economies of the *Leaders*, relatively efficient economies of the *Followers*, and relatively inefficient economies of the *Followers*. By introducing a target variable *Cluster Efficiency*, with domain of values (1, 2, 3, 4), we can identify each group of economies within each cluster and use the target variable in DT analysis to identify the split variables and their values that differentiate the groups. To identify the most meaningful splits, the decision maker can opt to display the resulting DT in the form of the easy-to-interpret decision rules and then concentrate on the rules that have a high probability for the occurrence of

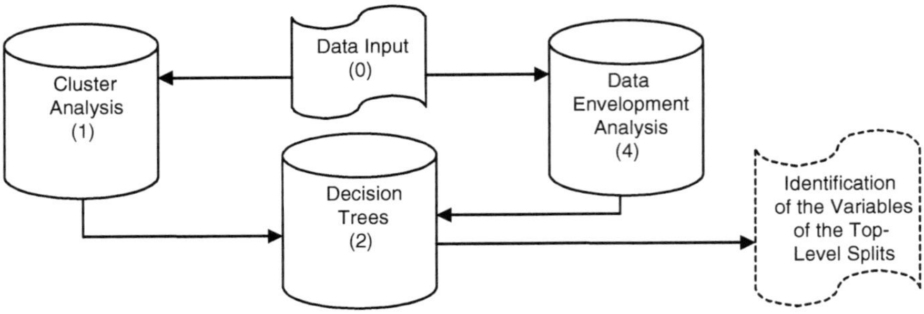

Fig. 5 What are some of the factors associated with the differences in relative efficiency?

the group based on the decision rule. Figure 5 illustrates Step 4, and the results are presented in Table 2.

The results demonstrate that the functionality provided by the DSS allows the decision maker to obtain important information regarding some of the factors that differentiate not only efficient and inefficient peers, but also efficient and inefficient non-peers. This information could be useful for the purposes of intra-group benchmarking, as well as for the purpose of formulating strategies for business units that are interested in intergroup transitioning.

Taken together, results of Step 4 allow the decision maker to obtain the following information regarding the heterogeneity of the sample in terms of the efficiency-based performance:

- Some of the variables associated with variation in efficiency are: *Productivity Ratio per Telecom Worker, Annual Telecom Investment, and Full-Time Telecommunication Staff %.*
- The peer group's variation in efficiency could be reduced via decreasing the levels of heterogeneity of such variables as:
  - *Productivity Ratio per Telecom Worker* and *Annual Telecom Investment* for Cluster1 (the *Leaders*)
  - *Full-Time Telecommunication Staff %* and *Productivity Ratio per Telecom Worker* for Cluster2 (the *Followers*)
- The non-peer group's variation in efficiency could be reduced via decreasing the levels of heterogeneity of such variables as: *Productivity Ratio per Telecom Worker* and *Full-Time Telecommunication Staff %.*

At this point, the decision maker is aware of the variables that are associated with the differences in regard to the efficiency-based performance; however, an additional benefit could be obtained from identifying complementarities between those variables that produce a synergistic effect on the output. The functionality of the DSS outlined in Step 5 allows the decision maker to identify some of the complementarities that may exist between the relevant to the production process variables.

Table 2 Decision rules generated by DT analysis

Group	Decision rule	Posterior probability
Group 1: efficient leaders	Productivity ratio per telecom worker ≥ 1.5674014075 & Annual telecom investment < $836,899,003 & Full-time telecommunication staff % ≥ 0.0039016912 & Annual telecom investment per worker ≥ $58	0.94
	Productivity ratio per telecom worker ≥ 4.1754445351 & Annual telecom investment per worker ≥ $58	1.00
Group 2: inefficient leaders	Annual telecom investment ≥ $836,899,003 & Full-time telecommunication staff % ≥ 0.0039016912 & Productivity ratio per telecom worker < 4.1754445351 & Annual telecom investment per worker ≥ $58	1.00
	Full-time telecommunication staff % < 0.0039016912 & Productivity ratio per telecom worker < 4.1754445351 & Annual telecom investment per worker ≥ $58	1.00
Group 3: efficient followers	Full-time telecommunication staff % < 0.0031414015 & Productivity ratio per telecom worker ≥ 3.8043909395 & GDP per Capita ≥ $519 & Annual telecom investment per worker < $33	1.00
	Total telecom services revenue ≥ 0.0118204323 & GDP per Capita < $519 & Annual telecom investment per worker < $33	1.00
Group 4: inefficient followers	Productivity ratio per telecom worker < 3.8043909395 & GDP per Capita ≥ $519 & Annual telecom investment per worker < $33	1.00
	Productivity ratio per telecom worker < 2.002357802 & $33 ≤ annual telecom investment per worker < $58	1.00

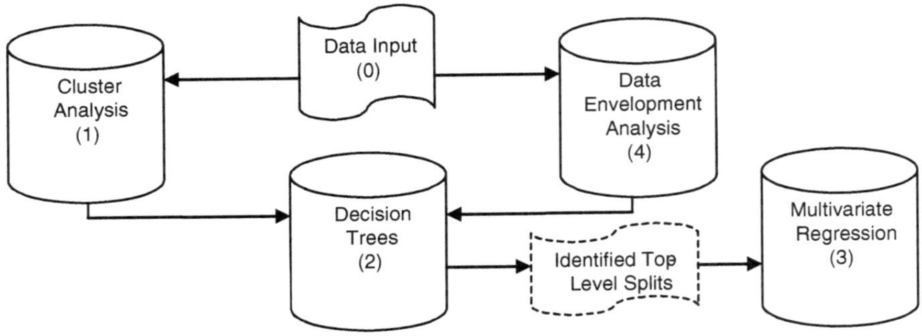

Fig. 6 What are some of the factors impacting the current levels of the relative efficiency?

Step 5 Are There Any Complementarities Between the Relevant Variables?

In order to identify existing complementarities between the production process variables, we construct a three-variable model consisting of two inputs—investments and labor, and one output—GDP (Fig. 6).

This will allow us to relate investments in telecoms, full-time telecom staff, and GDP as the following production function:

$$\text{GDP} = f(\text{investments in telecoms, full-time telecom staff}), \text{ represented as } Y = f(K,\ L).$$

This allows us to construct the following formulation to test for the presence of interaction:

$$\log Y = \beta_0 + \beta_1 \times \log K + \beta_2 \times \log L + \beta_3 \times \ \log K^2 + \beta_4 \times \ \log L^2 + \beta_5 \times \ \log K \times \log L + \xi,.$$

A test for the presence of the interaction between investments in telecoms and telecom staff would involve testing of the following hypothesis:

H0: $\beta 5$ is not statistically discernible from 0 at the given level of α.

The testing of the null hypothesis yielded the following results:

In the case of the *Leaders*, β estimate is 57.4954 (**p @ 95 %** is less than 0.0001)—this allows us to reject the null hypothesis of no interaction.

In the case of the *Followers*, β estimate is −2.1280 (**p @ 95 %** is 0.0087)—the null hypothesis cannot be rejected.

Overall, Step 5 allows the decision maker to conclude the following:

> Annual Telecom Investment and Full-Time Telecommunication Staff are complementary factors that may allow for improving the level of the efficiency-based performance of the organizational entities in the Leaders group (i.e. Cluster1).

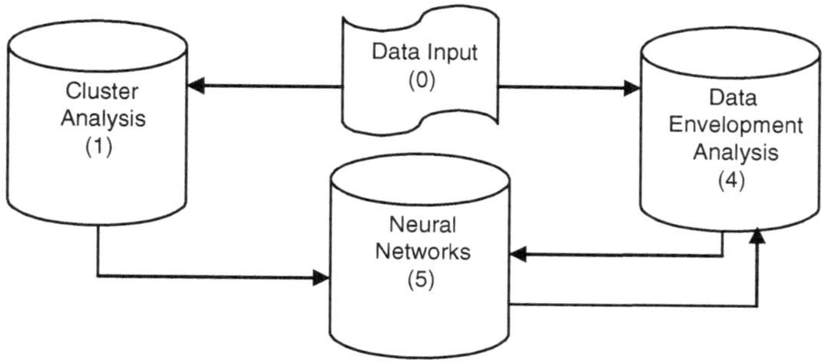

Fig. 7 What is the most effective way of increasing efficiency of the input–output process?

However, despite obtaining important insights regarding the presence of complementarities, the decision maker will still need additional information regarding the best route to improvements in the level of the efficiency of the production process, specifically as it relates to the production of outputs. For example, we determined that the *Leaders* are, on average, more efficient than the *Followers*; however, we also determined that the levels of investments and revenues of the *Leaders* are higher than those of the *Followers*.

This situation allows for two interpretations: first, members of the *Leaders* are more efficient than members of the *Followers* because of the superior process of conversion of inputs into outputs, and second, the members of the *Leaders* are more efficient because they have higher levels of inputs which allows for establishing and maintaining more efficient processes. Consequently, the design of DSS must allow for the functionality allowing for determining the most appropriate route to improvement in the production of outputs, namely whether to increase the level of inputs, or whether to improve the production processes first.

Step 6 What is a Better Way to Improve Production of Outputs?

Let us recall that NN analysis allows the decision maker to create a model of the input–output process in the form of an *input–output transformation* function, which then could be saved and applied to a new set of production inputs with the purpose of generating a new set of the production outputs (Fig. 7).

In the case of our illustrative example, NN analysis allows us to generate two transformation functions, *TF1* for the *Leaders* and *TF2* for the *Followers*. If we apply the inputs of the *Leaders* to TF2, we can simulate the level of outputs that members of the *Followers* would have produced if they had the levels of inputs of the *Leaders*. Conversely, if we apply TF1 to inputs of the *Followers*, we can simulate the level of outputs that members of the *Followers* would have obtained if they utilized process of conversion of inputs into outputs of the *Leaders*. Both simulated DEA models were created for the purposes of gaining insights into the most appropriate route of improving the level of efficiency of the *Followers*. Altogether, given 2 clusters, we end up with four DEA models:

DEA Model 1	Actual model based on the original inputs and outputs of the *Leaders*
DEA Model 2	Actual model based on the original of inputs and outputs of the *Followers*
DEA Model 3	Simulated model of the *Followers*, where the outputs are based on the inputs of the *Leaders*
DEA Model 4	Simulated model of the *Followers*, where the outputs are based on the process of input-output conversion of the *Leaders*

By comparing the scores produced by the original models with the scores of the simulated models, we can determine whether the *Followers* would get a greater gain in efficiency of production of outputs from increasing the level of inputs (DEA Model 3), or from improving the efficiency of conversion of inputs into outputs (DEA Model 4). Results of this comparison are summarized as follows:

- An increase in the levels of inputs of the *Followers* to the levels of the *Leaders* results in a decrease in the scores of the relative efficiency of the *Followers*.
- An improvement of the actual input–output transformation process of the *Followers* to that of the *Leaders* results in an increase in the scores of the relative efficiency of the *Followers*.

These findings allow a decision maker to determine that the economies of the *Followers*' group should not pursue an increase in the level of inputs as a means of increasing efficiency of output production; instead, improvements in the production process should serve as a means of increasing efficiency of production of output. Overall, the results of Step 6 offer evidence that:

> The existing inefficiencies of the Followers are associated with the inefficient processes of conversion of inputs into outputs, and not with the insufficient levels of inputs.

4 Conclusion

In this chapter, we presented a DEA-centric DSS that provides facilities for assessing and managing the relative performance of productivity-driven organizations that operate in unstable environments. The design of our DSS was guided by a set of requirements that are highly relevant to a productivity-driven organization's efforts to identify and evaluate multiple productivity models in order to select the most suitable one for the given organization. The resulting DSS is applicable to different organizational levels, including the country level and the firm level. In this chapter, we demonstrated the feasibility and usability of this DSS on country-level organizational entities.

Acknowledgments Material in this chapter previously appeared in: "Using Data Envelopment Analysis (DEA) for Monitoring Efficiency-Based Performance of Productivity-Driven Organizations: Design and Implementation of a Decision Support System," *Omega 41:1*, 131–142 (2013).

Reference

Piatkowski M (2003) Does ICT investment matter for output growth and labor productivity in transition economies? TIGER working paper series, No. 47., available on line at www.tiger.edu.pl

Chapter 14
Overview of the Value-Focused Thinking Methodology

Corlane Barclay

The chapter provides an overview of the value-focused thinking (VFT) methodology. Its main purpose is to introduce the reader to the major concepts of this methodology, particularly those that are relevant to the chapter that involves the use of VFT. It also discusses previous applications of the VFT methodology in information systems research.

1 Introduction

'The means to an end' is a common phrase where persons consider the sets of steps or objectives necessary to achieve an end goal. The common limitation of this, however, is the primary focus on the means and not on the end goals. Value-focused thinking (VFT) has the same precept; however, the key is the focus on the end while establishing a structured set of processes to assure that the alternatives and means are clear to achieving explicit values or end objectives. This chapter provides a general introduction to the VFT methodology, its potential benefits to information systems research, and examples of its applications in various contexts. VFT as the name suggests is value-centric where the values, i.e., the core set of considerations one cares about within a specific decision context. Therefore, it is centered on *what them is I want from the situation?* The core objectives and *how I am going to achieve it?* The means objectives or alternatives. Keeney (1994) argued that traditional thinking identifies alternatives and only after those are identified are the values considered. Such an approach is reactive, and therefore, the proactive value-centric approach considers the key activities that must occur to

C. Barclay (✉)
University of Technology: Jamaica, School of Computing and Information Technology, 237 Old Hope Road, Kingston, Jamaica
e-mail: clbarclay@gmail.com

K.-M. Osei-Bryson and O. Ngwenyama (eds.), *Advances in Research Methods for Information Systems Research*, Integrated Series in Information Systems 34,
DOI: 10.1007/978-1-4614-9463-8_14,

address the decision problem or context. The rationale for this approach is that it aids in making better decision since a more thorough understanding of the decisions and the context is achieved. Based on these principles, this methodology can be applied to any domain to help inform the stakeholders of the values and alternatives to facilitate improved decision making. Thus, this approach can be applied in any business or research situation.

2 Overview of Value-Focused Thinking (VFT) Methodology

Any decision situation should begin with values (Keeney 1996). According to Keeney (1996), the seminal authority on VFT, the approach describes and illustrates concepts and procedures for creating better alternatives for decision problems. Decision problems or situations are any set of concerns, issues, or unknowns within a defined context, such as a business. The determination of values is a key step in organizational strategic development process. Within a business or project context, clearly defined values and objectives may positively impact the performance of the project, as performance expectancy is aligned with the business or project values. The identification of these stakeholders' objectives will help to provide the business or project with a clear map to optimize performance. Unfortunately, current methods may lead to ambiguity and omission. The VFT provides an effective solution to this problem.

VFT is an alternative approach to identifying core objectives of organizational activities and procedures where emphasis is placed on the identification of values and then the alternatives in achieving these values. It is essentially about deciding what is important and how to achieve it, which encapsulates what the decision makers care about (Keeney 1992). As the name suggests, value is at the center of the decision-making process, values are fundamental and thus should be the driving force for our decision making (Keeney 1996). Values are principles used for the evaluation (Keeney 1992). Values that are of concern are made explicit by the identification of objectives. In this context, an objective is a statement of something that one desires to achieve (Keeney 1992). Values are seen as the desirability of a set of possible alternatives of consequences (Kenney 1994).

It was further explained that this paradigm is contrasted with the traditional approach, referred to as alternative-focused thinking (AFT) in three ways: (1) the effort in making the values explicit, this important step is done before other activities and these values are used to (2) identify decision opportunities, and (3) create alternatives. The conventional approach instead focuses on identifying alternatives, and then, the objectives or the values of the tasks follow (Keeney 1996). He contends that this traditional approach is a reactive approach and is a limited way to think through decision situations. While it is important to identify objectives, simply listing objectives is shallow as there is the need for greater depth, clear structure, and a sound conceptual base in developing objectives for strategic decision contexts (Keeney 1996).

2.1 The VFT Process

In explicating a clear structure in determining core objectives, the main steps in the VFT approach are identifying the objectives, structuring the objectives to distinguish means or end objectives, creating alternatives, identifying decision opportunities and developing the means-end network, Fig. 1. Some key underlying assumptions are that the key stakeholders relevant to the study and decision context are identified to gather the values and objectives important to them and that the steps although expressed as linear are iterative.

(1) *Develop an initial list of objectives and convert them into a common form* This step is concerned with determining what objectives are important to the stakeholders or decision makers and upon analysis convert responses to a common form. This process requires significant creativity and hard thinking. A useful approach is to engage in discussions about the decision situation, and ask, '*What would you like to achieve in this situation?*' (Keeney 1996). Three features are required in stating an objective explicitly—decision context, an object, and a direction of preference (Keeney 1996). This essentially means explicating the objective within its context based on the nature of the problem and determining exactly what the decision maker is ultimately trying to achieve. Objectives care classified into two types:

   (a) *Fundamental Objectives*—the end that the decision maker values in a specific decision context, and
   (b) *Means Objectives*—methods to achieve the ends (Keeney 1996, p 538) which are context dependent.

This implies that consideration of the particular nature or purpose will determine how the fundamental and mean objectives are formed.

Keeney (1994) identifies some techniques that are useful for identifying objectives. Developing a wish list by determining what the stakeholder wants or values, determining alternatives to an objective, considering goals, constraints or shortcomings to a decision context, for example, determining what areas require fixing, looking at generic objectives from different types of stakeholders, considering perspectives, and determining strategic or ultimate objectives are some of the strategies that can be employed to derive the objectives. The focus is on identifying, wants and needs without placing limitations of our thinking such as saying the particular objective is unreasonable or infeasible. Therefore, facilitating brainstorming from the stakeholders could be useful in developing an initial list.

Fig. 1 Key steps in the value-focused thinking process

Converting to common form simply means transferring all the sets of objectives into a generalized form that is representative of all the objectives identified. This step would require careful analysis of the responses from the stakeholders, taking account of how they express their wants and needs and putting all of these objectives together in a 'normalized' form. Therefore, there is the likelihood that the list of objectives will be reduced. It may be important to obtain verification from the stakeholders of the 'normalized' objectives before the next step to ensure that their views, wants, and needs are accurately represented.

(2) *Structure the objectives* Once the objective listing has been finalized, it becomes necessary to distinguish between those that facilitate the achievement of another (means) and those that are fundamental value of the stakeholder in a particular setting (end). In short, structuring the objective involves distinguishing the fundamental or end objectives from the means objectives or alternatives. As explained previously, fundamental objectives relate to the ultimate end objective while means objectives consider how to achieve those ends. It is expected that an initial list will include several variables, including mean and fundamental objectives. Asking additional questions on why the objectives are important will help to clarify the type of objectives while uncovering additional insights into the decision context.

The Why is it important? (WITI) test is the technique used to help distinguish between the fundamental and means objectives (Keeney 1994). For each objective, the WITI test is applied, and depending on the response, the type of objective is determined. Where the answer suggests that the objective is important is essential for this context and no other objectives are used as a basis for its importance, then the objective is at its end, i.e., end of fundamental objectives. Alternatively, where the answer suggests that the objective is important as a result of another objective, it means that objective is facilitating or an alternative to achieving the end, i.e., means objectives.

(3) *Create alternatives* New alternatives may be better able to achieve the fundamental objectives by using the same or fewer resources in a different way. The underlying principle is that alternatives should be created that best achieve the values specified for the decision situation (Keeney 1996). In essence, we need to critically assess the set of alternative means to achieving the fundamental objectives that go beyond the initial set that may come to mind. Keeney (1994) suggests that the range of alternatives persons tends to identify for any given decision context tend to be narrow. This provides a challenge in determining more robust decisions since the alternative narrow view may lead to ill-defined decisions or constrained thinking. In creating alternatives several approaches can be adopted (Keeney 1994). One way is to consider how to better achieve the fundamental objectives. This is done by focusing on one objective at a time, and considers desirable alternatives. After this step Keeney (1994) suggests that objectives are considered two at a time and identify alternatives that are desirable for both objectives. The step is followed at

subsequent stages by adding an objective to the set and continuing the process of analysis. The alternatives can be then refined into a common form. This may lead to conflicting alternatives but that discussion is outside of the focus of this chapter. Another approach is to use the means objectives to create additional alternatives.

(4) *Identify decision opportunities* Based on studies, this step is generally overlooked in the application of the VFT. Decision makers usually think of decision situations as problems to be solved, not as opportunities to be taken advantage of. Decision opportunities are those considerations to better achieve the overall values by formulating a decision situation (Keeney 1994). According to Keeney (1994) a decision opportunity is akin to prevention since it may minimize or avoid other decision problems.

Decision opportunities can be created in two ways by converting the existing a decision problem into decision opportunity or creating new decision opportunities. This process forces us to consistently think about better ways to get something done thereby creating opportunities. An example is changing your decision problem from whether to apply a certain research methodology to how to identify the approach that best answers the research questions posed for a given context.

(5) *The final step is building the means-ends objective network* The network is developed based on the structure of the objectives where the fundamental objectives are to the right and the means objectives leading to each other and the fundamental objectives are shown. The means-end network provides clarity on the interrelationships between the means objectives and how they influence the fundamental objectives. Aided with this tool, decision makers are better able to see the alignment of strategic business or project objectives. See Fig. 2.

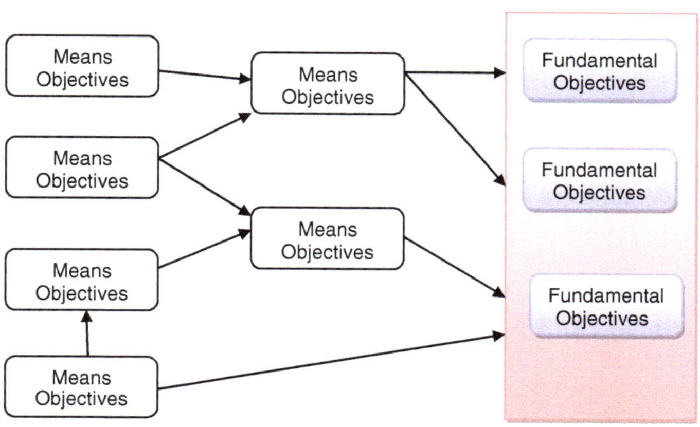

Fig. 2 Means-ends objective network diagram

2.2 An Illustration of the VFT

In the context of a graduate program, several stakeholders were questioned on the subject of maximizing graduate thesis outcome. The decision context is defined to be maximizing graduate thesis project outcome, which is about how to get the most value from this graduate exercise. Some examples of stakeholders are students, administrators, student advisors, and committee members.

To obtain a list of objectives, questions such as, *what do you value in this graduate thesis exercise? What do you expect from this graduate project?* were asked. From the list of objectives each participant was asked why each of their identified objectives was important. During this interview stage additional objectives may be identified or clarified. Some objectives identified were: to achieve graduate level content knowledge; obtain intellectual independence; conform to the ethical standards of the profession; and objectivity in the thesis review process.

In the determination of type of objectives, a student participant identified the need to improve research skills or enhance their understanding of research methodologies; however, in asking why these were important in completing their project, objectives such as the need to contribute or produce quality and relevant research and the ability to make a positive impact in the field were identified. Therefore, one can see the relationship between those two objectives. Assuming the latter to be the end objective after the interview stage, that is, the fundamental or end objectives with the means to achieve that being an understanding of research skills and methodologies, or achieving graduate content knowledge.

To identify alternatives, the fundamental objectives should be the central focus, thus asking Are there different or alternative ways to make a positive impact in the field? Objectives such as to volunteer on editorial committees develop patents, and other type of objectives may be identified. However, it is important to keep the decision context in mind. A continuous cycle of seeking alternatives to the objectives could be performed resulting in additional objectives and relationships. This activity supports the identifying decision opportunities stage of the VFT process by asking how a certain set of objectives could be best achieved. Once the exercise and process of objectives and alternatives are performed, the final step is graphically structuring the objectives into the means-ends objective network. The results should help show how to best achieve success in the graduate thesis project decision.

2.3 Benefits of the VFT Approach

The VFT approach can lead to better decisions. It is a proven methodology that has been used in various disciplines Sheng et al. (2005) including information systems and project management. This approach embodies a paradigm shift in thinking about how decisions can significantly improve decision making because values guide not only the creation of better alternatives but the identification of better decision situations (Keeney 1994). These better decision situations, which you create for yourself, should be thought of as decision opportunities, rather than as decision problems (Keeney 1996). Better decisions come about both because of insights provided by the thinking and because of specific procedures that view decisions through 'value-focused' glasses. Additionally, the approach provides a systematic, proven, and reliable way of identifying the relationships between objectives Sheng et al. (2005). In short, the approach facilitates structured guidance of decision situations through the identification of improved objectives and alternatives, improved understanding of the fundamental objectives of a decision context, and the structuring of objectives. As a proactive approach, it facilitates an active engagement in thinking about how a decision problem can be changed to a decision opportunity in effect determining how to perform or act better as a result of a situation or stimuli.

The key benefits of the VFT approach (Keeney 1992, pp 24–27) include (see also Fig. 3) as follows:

1. *Uncovering hidden objectives* Hard thinking facilitates going beyond the obvious to uncover objectives not previously considered. Additionally, obtaining perspectives from multiple decision makers enhance the chance of widening the breadth of objectives for a given decision situation of context.
2. *Guiding information collection* The technique guides information collection through the identification of objectives as decision makers are able to identify what they want or desire of a given context and explain why a given objective may be important.
3. *Improving communication* Where the opportunity exists to express the wants and needs such as in a business environment, communication and common understanding may be improved.
4. *Facilitating involvement in multiple-stakeholder decisions*. The nature of the technique and the process involved in identifying and structuring the objectives facilitates improved communication and engagement of multiple stakeholders in the decision process.
5. *Avoiding conflicting decisions* Conflicts are a normal part of multiple-stakeholder engagements; however, continued focus on the most desirable outcome and prioritization of objectives facilitate conflicting decisions. Further, the determination of the fundamental objectives guides the decisions and eliminates any conflicts or poorly defined decisions.
6. *Creating alternatives* Creating alternatives is about facilitating additional considerations or ways to achieve the end objectives. This effort provides an opportunity for identifying additional objectives and uncovers solutions or alternatives that will ultimately improve the decision process.
7. *Evaluating alternatives*. Once a wider set of alternatives is created, these alternatives can be examined and assessed to guide the stakeholder or decision maker in determining the best alternatives and thereby clear the path to a more effective decision-making process (better decisions).

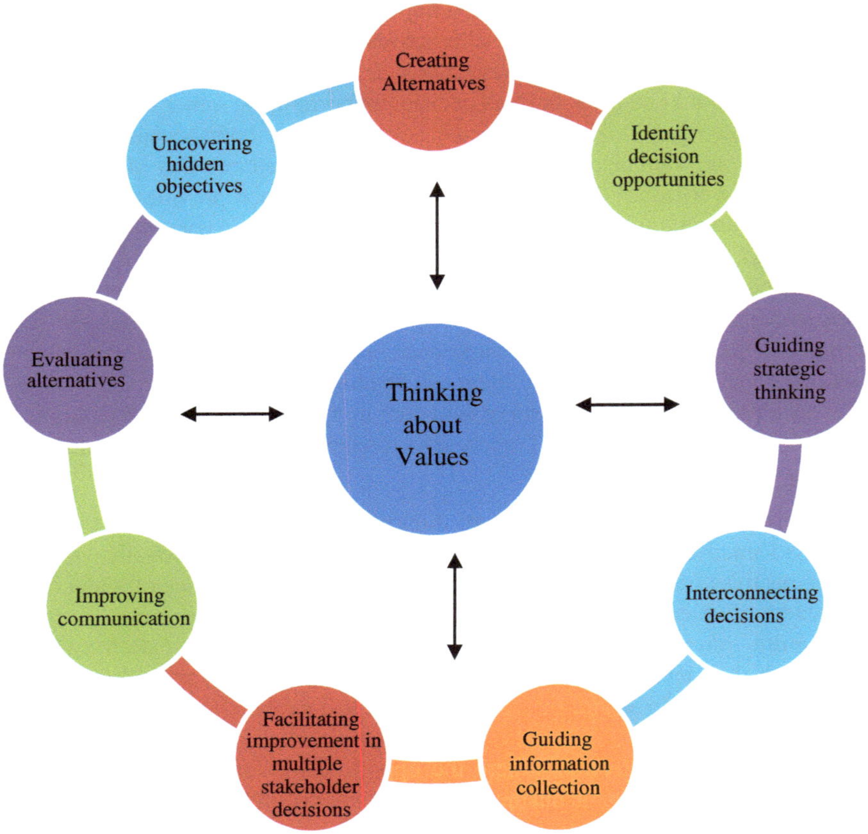

Fig. 3 Basis for quality decisions (Keeney 1994)

8. *Identifying decision opportunities* What can I do better? This question enables the decision maker to focus on opportunities and not be constrained by decision problems thereby possibly preventing future problems due to the attention placed on an unconstrained view of the decision situation.
9. *Guiding strategic thinking* Any opportunity to create improved objectives and an understanding of the decision situation will likely enhance strategic insight, a necessary characteristics of successful business. Therefore, a value-focused approach to solving any given problem or opportunity enables stronger decision support.

The benefits derived are supported by diverse applications. Orfelio (1999) performed a two-part study to investigate the differences between AFT and VFT among 58 psychology students found that the VFT approach provided a greater amount of objectives and more metrics for assessing these objectives. The second study proved that VFT assisted in the development of more robust alternatives.

Orfelio (1999) thus summarized that VFT is more complete, operational, and understandable compared with the traditional thinking approach.

This approach is not without its limitations, which can be seen as challenges mainly with the level of pervasiveness in the IS field and acceptance by practitioners, particularly in the current Jamaican environment. Some of the main challenges are as follows:

1. Difficult process that requires hard thinking to uncover objectives.
2. Most decision makers are accustomed to the traditional approach to decision making thus initial buy-in may be challenging.
3. Its adoption in the IS field is emerging and is relatively sparse.

Despite its challenges, it is apparent that this approach may have significance in identifying key organizational considerations within different contexts. Examples of its application are in diverse areas outside of the information systems (IS) domain, such as tourism management (Kajanus et al. 2004), environmental risk considerations (Gregory et al. 2001), improve watersheds (Merrick and Garcia 2004), and to select simulation tool for the acquisition of infantry soldier systems (Boylan et al. 2006).

3 Examples of VFT Applications in the IS Domain

Based on the multiple studies, the application of VFT is beneficial with its ability to uncover important objectives that may not have been initially considered. Armed with this benefit, decision makers are able to identify core objectives, how to achieve them and monitor progress through different stages. Though its application in the IS field has been sparse, its track record in other fields, and its proven benefits provide an insightful perspective to pursue. Examples of VFT applications in information security, project management evaluation, mobile application, Internet/e-commerce are displayed in Table 1. The literature underlined the diverse applicability of VFT to support decisions.

3.1 Information Security

The VFT approach was used to assess the value of information security in organizations from the perspectives of key members of the organization (Dhillon and Torkzadeh 2006). They used 103 managers to identify what they considered to be key values or considerations in managing IS security. The results were validated using seven IS security experts. The results implied that the maintenance of IS security requires organizational considerations to augment the technical component (Dhillon and Torkzadeh 2006). With the need to identify key areas of concern for security planning, the level of ICT security awareness was explored to help identify key areas of concern to address in ICT security awareness programs

Table 1 Application of VFT to IS domain

Domain	Description
Information security	• To assess the value of information security in organizations from the perspectives of key members of the organization (Dhillon and Torkzadeh 2006) • To identify key areas of concern to address in ICT security awareness programmes in an academic environment (Drevin et al. 2004)
Internet/e-commerce	• To examine the strategic implications of mobile technology in a leading publishing company (Sheng et al. 2004). • To determine the value of mobile applications in a utility company (Nah et al. 2005) • To identify the values of silent commerce Sheng et al. (2005) • To identify the factors influencing trust in mobile commerce and to explain the development of such trust using a means-ends objective network (Siau et al. 2003) • To measure the factors that influence Internet commerce success (Torkzadeh and Dhillon 2002) • To assess individual privacy concerns for Internet commerce (Dhillon et al. 2002) • To understand the value of Internet commerce (Chang et al. 2004) • The value of Internet commerce to the customer (Keeney 1999)
Education	• To understand the values of mobile technology in education (Sheng et al. 2010)
Electronic government	• To determine how citizens value e-government services (Park 2008)
Project management	• To aid in for developing performance criteria & measures for information systems (IS) projects (Barclay and Osei-Bryson 2010) • To help to measure the performance of information systems (IS) projects (Barclay and Osei-Bryson 2009) • To understand the stakeholders values of project success (Barclay and Osei-Bryson 2009a) • To help measure the strategic contributions of programs (Barclay and Osei-Bryson 2009b)
Business process management	To determine the values of business process (Neiger and Churilov 2004)

(Drevin et al. 2007). The research confirmed that the objectives were aligned to general security goals combined with additional findings linked to social and management issues (Drevin et al. 2007).

3.2 Value of Internet/E-Commerce

The value of m-commerce was investigated using the work system framework and the VFT approach to help understand the values most important to m-commerce customers (Siau et al. 2003). As expected, based on previous investigation, the study revealed some important issues and concerns of these customers (Siau et al. 2003), which may have been ignored otherwise. Trust in m-commerce and e-commerce, essential issues in the current environment, was also evaluated to determine the important objectives and how they may impact m-commerce and e-commerce activities (Chen and Dhillon 2003; Siau et al. 2003). The identification of these critical objectives allows decision makers to be equipped with the knowledge of trust characteristics to facilitate successful commerce activities.

With the advancement of mobile technology, it has shown value potential for organizational use. The strategic value of mobile technology on organizations was investigated to obtain better understanding through the examination of a leading publishing company through the VFT lens Sheng et al. (2005). The use of the VFT approach provided additional insights into the strategic implications of using mobile technology to support organizational units Sheng et al. (2005). Research on mobile technologies was extended to its applicability in a public utility company to discern the values associated with its use (Nah et al. 2005).

Keeney (1999) used the VFT approach to help determine the value proposition of Internet commerce to customers. Over 100 individuals were interviewed to identify the objectives they deemed important. The participants were interviewed in groups or individually, and values were identified and discussed in an iterative manner. A sample of the participants was later used to prioritize these objectives. The results of the fundamental and mean objectives were able to pinpoint the main concerns and values of the customer. The design of future Internet commerce systems, creation, and redesign of products that are aligned with customers' values are some of the potential use of the findings (Keeney 1999).

Torkzadeh and Dhillon (2002) also used VFT as a platform to investigate the factors that influence the success of Internet commerce because efforts to develop measures have been hampered by the rapid development and use of Internet technologies and the lack of conceptual bases necessary to develop success measures. They designed two instruments to measure the mean and fundamental objectives, i.e., the means objectives that influence online purchase (e.g., Internet vendor trust) and the fundamental objectives that customers perceive to be important for Internet commerce (e.g., Internet product value). Their conclusions provide the basis for decision makers to help distinguish between effective and ineffective Internet sites. Dhillon et al. (2002) continued similar work in this area with the assessment of individual privacy concerns for Internet commerce.

3.3 Project Management Performance

The project objectives measurement model (POMM) is grounded on several principles of the value-focused thinking (VFT) and goal question metric (GQM) techniques (Barclay and Osei-Bryson 2008). The evaluation of the proposed model was performed in two parts: a team of industry experts examined the principles of to model and provided feedback on its practicability to practice, and a case study of a Caribbean educational institution's IS graduate program development was used to illustrate the procedures of the model. The study attempts to address some of these concerns through the development of project measures that are aligned to key project stakeholders' values and objectives within the unique project contexts. It is argued that objectives are the key performance criteria of the project, and hence, measures must be aligned to these criteria, and formal procedures should be in place to assure that these objectives and measures are carefully developed and reflective of the persons to whom the project matters, the stakeholders.

A formal method to develop a comprehensive set of performance criteria or objectives to aid in the management of project and guide its performance evaluation was developed (Barclay and Osei-Bryson 2010). The project performance development framework (PPDF) primarily relies on the principles and advantages of VFT to elicit performance criteria based on the values of the stakeholders in the project. The effectiveness of the model was illustrated through the use of three project cases in different environments. The technique was also used in the context of developing a measurement framework for programs of projects (Barclay and Osei-Bryson 2009b) and to understand stakeholders' values pertaining to project success (Barclay and Osei-Bryson 2009b).

The technique has been applied in other areas such as education in the context of the applicability of mobile technology in that environment (Sheng et al. 2010), the value of e-government services (Park 2008), and business process management (Neiger and Churilov 2004).

4 Chapter Summary

Value-focused thinking (VFT) provides multiple benefits to aiding the decision-making process and resonates as a viable technique for information systems research. Based on the discussion, there is a clear opportunity to extend the application of VFT in multiple areas of IS such as cyber-security, knowledge management, data mining, and education. Further, there is an opportunity to further explicate and extend the principles of VFT.

The VFT approach is a value-centric approach that facilitates identifying fundamental objectives or values to a decision context to enhance the decision-making process. It is therefore applicable in decision support situations. The steps include identifying the set of objectives based on asking the stakeholders or decision makers what is important to them, and reducing the list of objectives to a common form, the objectives are then structured into means and fundamental objectives, creating alternatives, and decision opportunities are also utilized to refine the process, and a means-end network is then developed to reflect the set of values and alternatives important in the specific decision situation. The technique provides multiple benefits including uncovering hidden objectives, improving communication and engaging stakeholders, and creating and evaluating alternatives that guide the decision process. Therefore, there is value in using the VFT technique in contexts that require better decisions that will inform better thinking about a decision situation.

References

Barclay C, Osei-Bryson K-M (2008) The project objectives measurement model (POMM): an alternative view to information systems project measurement. Electron J Inf Syst Eval 11(3):139–154

Barclay C, Osei-Bryson KM (2009a) Determining the contribution of IS projects: an approach to measure performance. In: System sciences, 2009. HICSS'09. 42nd Hawaii International conference on. IEEE, pp 1–10

Barclay C, Osei-Bryson K-M (2009b) Toward a more practical approach to evaluating programs: the multi-objective realization approach. Project Manage J 40(4):74–93

Barclay C, Osei-Bryson KM (2010) Project performance development framework: an approach for developing performance criteria and measures for information systems (IS) projects. Int J Prod Econ 124(1):272–292

Boylan GL, Tollefson MES, Kwinn LCMJ, Guckert RR (2006) Using value-focused thinking to select a simulation tool for the acquisition of infantry soldier systems. Syst Eng 9(3):199–212

Chang JCJ, Torkzadeh G, Dhillon G (2004) Re-examining the measurement models of success for Internet commerce. Inf Manage 41(5):577–584

Chen SC, & Dhillon GS (2003) Interpreting dimensions of consumer trust in e-commerce. Inf Tech Manage 4(2-3):303–318.

Dhillon G, Torkzadeh G (2006) Value-focused assessment of information system security in organizations. Inf Syst J 16(3):293–31

Dhillon G, Bardacino J, Hackney R (2002) Value focused assessment of individual privacy concerns for internet commerce. ICIS 2002 Proceedings. Paper 67. http://aisel.aisnet.org/icis2002/67

Drevin L, Kruger HA, Steyn T (2007) Value-focused assessment of ICT security awareness in an academic environment. Comput Secur 26(1):36–43

Gregory R, Arvai J, McDaniels T (2001) Value-focused thinking for environmental risk consultations. Res Soc Prob Public Policy 9:249–273

Kajanus M, Kangas J, Kurttila M (2004) The use of value focused thinking and the A'WOT hybrid method in tourism management. Tourism Manage 25(4):499–506

Keeney RL (1992) Value-focused thinking. Cambridge, Massachusetts: Harvard University Press

Keeney RL (1994) Creativity in decision making with value-focused thinking. Sloan Manage Rev 35:33–33

Keeney RL (1996) Value-focused thinking: Identifying decision opportunities and creating alternatives. Eur J Oper Res 92(3):537–549

Keeney RL (1999) The value of Internet commerce to the customer. Manage Sci 45(4):533–542

Merrick JR, Garcia MW (2004) Using value-focused thinking to improve watersheds. J Am Plann Assoc 70(3):313–327

Nah F, Siau K, Sheng H (2005) The value of mobile applications: a utility company study. Commun ACM 48(2):85–90

Neiger D, Churilov L (2004) Goal-oriented business process modeling with EPCs and value-focused thinking. In: Business process management. Springer, Berlin, pp 98–115

Orfelio G (1999) Value-focused thinking versus alternative-focused thinking: effects on generation of objectives. Organ Behav Hum Decis Process 80(3):213–227

Park R (2008) Measuring factors that influence the success of E-government initiatives. In: Hawaii international conference on system sciences, proceedings of the 41st annual. IEEE, pp 218–218

Sheng H, Nah FFH, Siau K (2005a) Strategic implications of mobile technology: A case study using value-focused thinking. J Strateg Inf Syst 14(3):269–290

Sheng H, Nah Fiona F-H, Siau K (2005) Values of silent commerce: a study using value-focused thinking approach. AMCIS 2005 Proceedings. Paper 192 http://aisel.aisnet.org/amcis2005/192

Sheng H, Siau K, Nah FFH (2010) Understanding the values of mobile technology in education: a value-focused thinking approach. ACM SIGMIS Database 41(2):25–44

Siau K, Sheng H, Nah F (2003) Development of a framework for trust in mobile commerce. In: Proceedings of the second annual workshop on HCI research in MIS, pp 85–89

Torkzadeh G, Dhillon G (2002) Measuring factors that influence the success of Internet commerce. Inf Syst Res 13(2):187–204

Chapter 15
A Hybrid VFT-GQM Method for Developing Performance Criteria and Measures

Corlane Barclay and Kweku-Muata Osei-Bryson

This chapter utilizes the principles of the Value Focused Thinking (VFT) and Goal Question Metric (GQM) to present a tool for managing the foundation processes of projects, programs and portfolios. The Project Performance Development Framework (PPDF) promotes the use of formal steps for identifying project criteria and measures that are used as the basis to evaluate the performance of the project. It addresses critical problems in project management where objectives typically are not clearly defined and stakeholders have multiple views and objectives on what is important to them in the project, which in turn impacts how the project performance is perceived. The PPDF has several iterative steps including stakeholder identification, objectives identification and structuring, definition of project measurement and prioritization to aid in stakeholder management and project objectives identification, prioritization and management. The PPDF artifact presented is demonstrated and evaluated for utility through a single Process Quality Management Development Project. The findings support the goals of PPDF and show that the formal approach provides added advantages in identifying important values of the stakeholders thereby improving the likelihood of managing stakeholders' needs and achieving identified goals.

C. Barclay (✉)
School of Computing and Information Technology, University of Technology, 237 Old Hope Road, Kingston, Jamaica
e-mail: clbarclay@gmail.com

K.-M. Osei-Bryson
Department of Information Systems, Virginia Commonwealth University, 301 W. Main Street, Richmond, VA 23284, USA
e-mail: kmosei@vcu.edu

K.-M. Osei-Bryson and O. Ngwenyama (eds.), *Advances in Research Methods for Information Systems Research*, Integrated Series in Information Systems 34,
DOI: 10.1007/978-1-4614-9463-8_15,

1 Introduction

The chapter presents the Project Performance Development Framework (PPDF) for developing performance criteria and measures for the different forms of projects, i.e. project, programs and portfolios. The development of PPDF involves the adoption and adaption of previously proposed techniques. It involves the principles and techniques of stakeholder identification, Value Focused Thinking (VFT) methodology and Goal Question Metric (GQM) method. The stakeholder identification methods of various researchers (Lyytinen and Hirschheim 1988; Sharp et al. 1999) are adapted to produce a stakeholder identification method, which has been noted as a non-trivial issue (Pouloudi and Whitley 1997; Sharp et al. 1999); the VFT methodology (Keeney 1992) is used to elicit and structure project objectives from the key stakeholders; and the GQM method (Basili and Weiss 1984) to derive measures that are linked to the identified project objectives. PPDF is flexible with regards to prioritization, and can accommodate various techniques including the Analytic Hierarchy Process (AHP) (Saaty 1990), the Qualitative Discriminant Process (Bryson 1997).

Technology (IS/IT) projects have continued to have a paradoxical presence in contemporary organizations; they play an important, often strategic role yet are often perceived as troubled ventures (KPMG 2005). Within the last decade, there have been increasing calls to improve project competencies and capabilities of IS/IT projects to help overcome some of its challenges (Morris 2004; Winter et al. 2006b). Given that executives and other project stakeholders have a vested interest in assessing the real value that these projects provide to organizations (Brynjolfsson 1993; Melville et al. 2004), there is the need for appropriate measurement and support tools for more effective monitoring and controlling of these projects (Hartman and Ashrafi 2002; Thamhain 1994). It is reasonable to assert that not only projects will be impacted but all other forms of project activities, including programs of projects and portfolios of projects. In this regard there are three interconnected problems associated with the assessment of the performance of IS/IT projects, programs and portfolios that are the focus of this research:

1. The varied perception of performance.

Stakeholders have a vested interest in the activities and outcome of the project (PMI 2004), and success may mean different things to them which often result in different perceptions of success and how the project performed (Agarwal and Rathod 2006; Shenhar et al. 2001). This has significant implications as the project can be deemed a complete failure by one group and success by another.

2. Unclear and incomplete objectives have been shown to contribute to the perception of failure (Ewusi-Mensah 1997; The Standish Group 2001, 2004):

Objectives are the criteria by which the performance is based (PMI 2004), thus is it important to get this information right through the reflection of the project stakeholders and the integrated dimensions of project performance, i.e. the project process, project management process and the product (Baccarini 1999; Barclay 2008). An analysis of current academic literature relating to project evaluation shows that limited attention is shown to structured procedures to help identify or elicit these objectives from the stakeholders, outside of product requirements (CMMI 2002; Cohen and Graham 2001), or high or abstract level guidance that is not easily implementable (Atkinson 1999; Kaplan and Norton 1992).

3. The traditional system of project measures (i.e. triple constraints: cost, time, quality) that are commonly used to assess the performance of IS projects, have been identified as being incomplete (Atkinson 1999; Atkinson et al. 2006; Cohen and Graham 2001; Wateridge 1998) yet have dominated practice (Agarwal and Rathod 2006; KPMG 2005) and have been used as the basis to suggest high incidences of failure (Glass 2005).

Moreover, researchers and practitioners have argued that the triple constraint framework is an incomplete measurement framework because it does not sufficiently reflect the realities of contemporary projects (Atkinson 1999; Atkinson et al. 2006; Cohen and Graham 2001; Wateridge 1998). Cohen and Graham (2001) argued that a fundamental shift is needed to highlight satisfying customers, managing cash flow and shareholder value, selecting best time to market and other considerations rather than fixed specification, budget and fixed deadline. Hartman and Ashrafi (2002) noted that with changing business conditions traditional project performance measures are no longer effective for the evaluation of contemporary projects. Further, performance measures should be used not only for monitoring but as a basis for learning and improvement (Wegelius-Lehtonen 2001). Therefore, with continued sole reliance on the triple constraint paradigm there is a risk that other relevant and important project objectives are not accounted for and evaluated during the project such as project learning or use related objectives, resulting in learning opportunities being missed. It is noted that if practice is to persist with these trends, IS projects may continue to struggle to justify its contributions and learn from the past.

Against this background the research attempts to answer the following questions: *How can practitioners account for and identify performance criteria that are representative of the perspectives of the stakeholders* and *how do practitioners identify suitable measures for these performance criteria*. This study proposes a framework for the elicitation and development of objectives and measures that are relevant to the project from the perspectives of its stakeholders. Our PPDF provides a formal methodology for:

1. Identification of project stakeholders.
2. Elicitation and structuring of project objectives based on perspectives of the relevant stakeholders.
3. Prioritization of these project objectives.
4. Elicitation and identification of measures that could be used to evaluate each project.
5. Analysis of the suitability of these objectives and measures by stakeholders.

It is important to underline that the definition or development of performance criteria is distinct from the act of measurement. Hence, the research focus is to introduce a cogent method to identify and develop sound performance measurements which can be aligned and integrated with the organisation's measurement framework. Within this context, an evaluation process includes the commitment of a particular evaluation strategy or framework accompanied by a method to elicit criteria and measures from the relevant stakeholders. The research focuses on the latter, and posits that before an organization or its projects can effectively apply an evaluation method they need to be assured that they are evaluating or tracking the *right* criteria and measures.

The research involves a design science approach, and follows the guidelines prescribed by Hevner et al. (2004). Design research paradigm is fundamentally a problem solving paradigm (March and Smith 1995) where it involves the creation and evaluation of artefacts intended to solve identified organizational problems (Hevner et al. 2004). The artefacts can be categorised into *constructs, models, methods, instantiations* and *better theories* (March and Smith 1995). Thus, a suitable framework for developing and aligning project performance criteria and measures that could support the management and evaluation processes of IS projects is proposed and evaluated. It should be noted that the proposed framework is also applicable to project management endeavours beyond the IS projects. For example scheduling and supply chain management (Ala-Risku and Kärkkäinen 2006; Arshinder et al. 2008) in many cases involve project management and evaluation activities. An observational case study approach is used to illustrate and evaluate the applicability of the PPDF. The main expected contribution of this research is the design artefact aimed at improving current project evaluation and support processes; the provision of additional support to the measurement framework that is adopted by an organization or project; and an enhancement of the organizational project competencies through more effective identification and analysis of stakeholders' needs, values and strong or weak performance areas.

2 Research Background

2.1 Project and Project Performance Evaluation

Projects are temporary and unique activities that result in a new product, service or results (APM 2006; PMI 2004). Project performance is seen as an achievement of the *project management (PM) success, project success* and *product success* and are aligned to the performance criteria of the individual stakeholders (Barclay 2008; Collins and Baccarini 2004), Fig. 1 in Chap. 5. Project management (PM) success considers the achievement of the PM related objectives, inclusive of the traditional measures, i.e. time, cost and quality (PMI 2004). Project success considers the objectives of the project stakeholders (Collins and Baccarini 2004;

Cooke-Davies, 2002). Product success considers the final product or outcome of the project and stakeholders' product satisfaction (Cooke-Davies 2002) such as the acceptance of a developed application by the client. Taking into consideration all these dimensions can provide for improved analysis of the performance of IS projects as the focus is shifted from a single dimension to multiple dimensions suitable for multiple settings.

The performance management literature has provided guidance in the type of evaluation strategies that can be applied in practice. In this context, performance measurement is the measurement and monitoring of the project's performance criteria as defined by the stakeholders and representative of the project performance dimensions. Some key recommendations for Performance Measurement Systems (PMS) include considerations for the contexts and stakeholders. The PMS should allow for members of [the project] and organization to understand how the measures affect the project, and organization, (Dixon et al. 1990) which ties into the strategic significance of the particular activity. Thus the strategic role of the company [or purpose of the project] should be understood by the stakeholders (Azzone et al. 1991; Dixon et al. 1990). Multi-criteria should be used to evaluate group activities and the measurement should be easy to understand by the stakeholders (Dixon et al. 1990) and contain measurable and well-defined criteria for the organization and project (Globerson 1985). Folan and Browne (2005) also recommended that PMS should have two components: a structural element to facilitate management and selection of appropriate performance measures and procedural element that provides guidelines to how the performance measurement should be carried out. Despite several approaches being observed in practice, these are usually a collection of best practices and PMS, with varying degrees of success (Folan and Browne 2005).

Researchers have also recommended that objectives and measures beyond time, cost and specifications are necessary for effectively assessing performance. The *critical success factors* stream has focused on identifying environmental, organizational and project conditions that are critical to help to assure better performance, including top management support, commitment and technical know-how of project teams (e.g. Belassi and Tukel 1996; Bryde and Robinson 2005; Fortune and White 2006; Freeman and Beale 1992; Lim and Mohamed 1999; Milis and Mercken 2002; Shenhar et al. 2001, 2002). The *business value contribution* stream has considered approaches to effectively assess the value of the IT investments (e.g. Fitzgerald 1998; Kaplan and Norton 1992; Kumar 2003). The *critical success criteria* stream proposed methods or framework under which the measurement practices of IT/IS projects can be guided (e.g. Atkinson 1999; Barclay 2008; Nelson 2005). Some of the methods include development of collaborative strategy maps and measures for the automobile industry (Niebecker et al. 2008), performance scorecards for evaluating IS project activities (Barclay 2008; Martinsons et al. 1999), the provision of structured framework for the lifecycle management of IT projects that span the selection, implementation and evaluation phases (Stewart 2008). Accordingly, tangible and intangible objectives such as financial rewards, operational efficiencies improvements, learning opportunities

Table 1 Summary of literature

Key categorization	Project performance criteria	Literature
Project management and project team	Minimization of project cost and project duration; strong project commitment; communication; planning; conformance to budget, time, scope requirements; project functionality, project efficiency, management involvement	Atkinson (1999), Bryde and Robinson (2005), Freeman and Beale (1992), Nelson (2005), Shenhar et al. (2001, 2002)
Client and other stakeholders	Satisfaction; endorsement; acceptance; user involvement; utility; use; safety; impact on customer; customer service; increased responsiveness	Belassi and Tukel (1996), Bryde and Robinson (2005), Kumar (2003), Lim et al. (1999), Nelson (2005)
Project product or service	New product or market; safety; commercial performance; technical performance business and direct success; financial rewards; implementability; flexibility	Barclay (2008), Lim et al. (1999), Fitzgerald (1998), Freeman and Beale (1992)
Preparation for the future	Value, personal growth, learning, readiness for the future	Barclay (2008), Nelson (2005), Freeman and Beale (1992), Kaplan and Norton (1992)

and enhancements to knowledge about managing the project or about the project's outcome are some of the key indicators that should also be tracked and evaluated. A summary of some of the relevant literature and performance criteria is highlighted in Table 1.

2.2 *Accounting for Stakeholders' Objectives*

Stakeholders refer to individuals and groups, internal and external to the organization, that are most involved in a project with a vested interest in its outcome or contribution (PMI 2004). They may include contractor, sponsor, project team and client (PMI 2004). These stakeholders may have certain expectations and consequently engage in behaviour that may be constructive or destructive (Bourne and Walker 2006). While stakeholders may have similar or divergent views on what is important in a project it is essential to manage and account for their perspectives, as best as possible, in accounting for how the project performed. Given that an incomplete set of stakeholders may result in an incomplete set of project objectives and criteria that can result in challenges for the project PMI (2004), it is important to identify all significant stakeholders and to elicit their values and objectives. While various studies (Pouloudi and Whitley 1997; Sharp et al. 1999) have recognised that the identification of stakeholders is an important activity, the need for guidance for this step has often not been recognised in practice, particularly with regards to the management of IS projects.

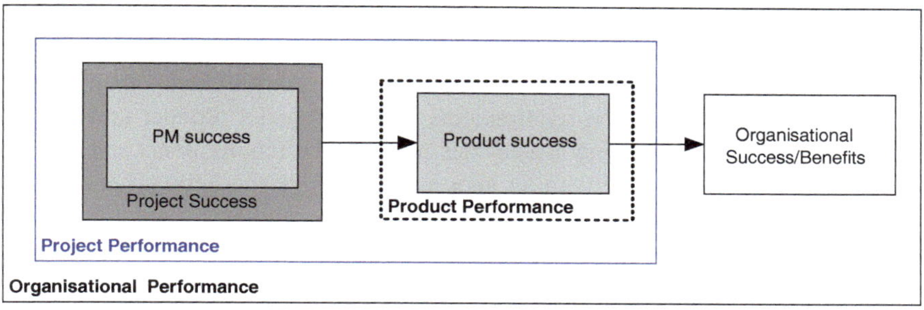

Fig. 1 Project performance constituents

Pouloudi and Whitley (1997) took an interpretive stance in proposing four principles in how stakeholders are perceived, which they used to identify stakeholders in the drug use management domain: (1) stakeholders depend on the specific context and time frame, thus the map of stakeholders should reflect the domain and be reviewed over time; (2) the stakeholders cannot be viewed in isolation as each stakeholder identified may lead to others; (3) the position of each stakeholder may change over time and can explain the past and plan for the future; and (4) feasible options may differ from the stakeholders' wishes hence exploration of political, economic and other issues are necessary. Sharp et al. (1999) focused on the level of interactions between stakeholders which include a set of baseline or core stakeholders followed by satellite stakeholders. Lyytinen and Hirschheim (1988) proposed that the nature of the IS; the type of relationship of the stakeholder to the IS; the direct or indirect 'depth of impact' in the identification; and the level of aggregation influence the stakeholder group.

2.3 Eliciting and Structuring Objectives

Project objectives are measurable success or performance criteria of the project (PMI 2004), and may include project process objectives, project management objectives and product objectives which are aligned to the project performance dimensions (Fig. 1). Therefore, it is imperative to consider the full context of the project's performance dimensions as viewed through the lenses of the project's stakeholders for the lack of clearly defined goals and objectives have an impact on IS project failure (Ewusi-Mensah 1997; The Standish Group 2001, 2004) given that "what you measure is what you get" (Kaplan and Norton 1992). Further clearly defined project objectives provide a basis for clear communication, maintaining focus on achieving desired outcome(s) and building commitment by allowing small gains toward larger ends (Katzenbach and Smith 2003). Also as suggested by Cannon-Bowers et al. (1993), a high level of performance requires that the stakeholders have common understanding of the group's objectives, their own responsibilities toward those objectives, and the procedures they will use to attain these objectives.

Gathering project objectives can be considered akin to gathering requirements. Requirements management techniques have been used to define the project's product objectives. CMMI for example, describes requirements management as the process to manage the requirements of the project's product and product components and to identify inconsistencies between the requirements and the project's plan (CMMI 2002). Emerging approaches include the agile strategy, these approaches are mainly used to manage software requirements and emphasise flexible and progressive elaboration of objectives of the users and other stakeholders over the life of the project (Alleman 2005). A challenge however is that there is no clearly defined method to ensure that the objectives reflect the purpose of the project and what the stakeholders expect from the project, above and beyond the project's product.

2.4 Overview on the Value Focused Thinking Method

The VFT methodology of Keeney (1992) aims to uncover hidden strategic objectives of diverse managerial processes (Keeney 1992, 1996). It has been used successfully in several areas including operations management discipline (Keeney 1992, 1996). Within the IS domain several examples of investigation are noted: trust in m-commerce and ecommerce and how they may impact m-commerce and ecommerce activities (Chen and Dhillon 2003; Siau et al. 2004); the value of information security in organizations from the perspectives of key members of the organization (Dhillon and Torkzadeh 2006). Interestingly, there is no known application in the IS project management domain or in the investigation of IS performance measurements.

The application of the VFT technique includes some important steps (Keeney 1992):

1. ***The Identification of Key Stakeholders***. As project boundaries extend beyond the organization (Engwall 2003) stakeholders' groups will also extend. While it may be impractical to obtain values from all stakeholders, having an appreciation of who are the stakeholders in the project and who play an essential role in the project is the logical first step.
2. ***The identification of Stakeholders' Values***. Values are those principles that encapsulate what a person cares about or values in a specific situation (Keeney 1992, 1996). Therefore, taking a structured approach to understanding the values of key stakeholders at the start of the project may allow for improved monitoring and control throughout the project.
3. ***Converting Values to Objectives***. An objective is characterised by three features: *a decision context, an object and a direction of preference* (Keeney 1996). This essentially means explicating the objective within its context based on the nature of the problem and determining exactly what the stakeholder is ultimately trying to achieve. It is also possible to derive more than one objective

from a specific value statement (Drevin et al. 2007). This further underscores the importance of interacting with the stakeholders to better understand their vision, values. Objectives are classified into two types: (1) *fundamental objectives*—the end that the decision-maker values in a specific decision context, and (2) *mean objectives*—the methods to achieve the ends (Keeney 1996 p. 538) which are context dependent. This implies that consideration of the particular nature or purpose will determine how the fundamental and mean objectives are formed.

4. ***Determining the Relationships between Objectives.*** To perform this step, asking why each objective is important will help to distinguish between fundamental and means objectives (Keeney 1992, 1996) which results in the means-end objective network (Keeney 1992, 1996). Aided with this tool, decision makers are better able to see the alignment of strategic objectives or project objectives.

2.5 Eliciting Measures: Overview on Goal-Question-Metric Method

It is necessary to have the project measures reflective of the stakeholders' objectives to facilitate a consistent evaluation process. The GQM is used to support this step, it is a metric generation technique that develops performance metrics aligned with each goal (Basili and Weiss 1984), or within this context, objectives of particular process e.g. delivery of software to an organization. It utilises a top-down method for the identification of metrics needed for certain goals by asking questions linked to these goals, the major steps of which are:

1. Formalise measurement goals
2. Identify quantifiable questions
3. Define the measures to be used
4. Prepare plan for implementing and interpreting the measures.

GQM is based on two principal assumptions (Basili 1999): a measurement program should not be 'metrics-based' but 'goal-based', and the definition of goals and measures needs to be tailored to the individual organization. It has also been applied in several evaluation exercises (Caldiera 1994; Basili and Weiss 1984; Esteves et al. 2003) to strengthen the measurement process. Applying these principles into the IS project evaluation context can strengthen the generation of stronger metrics associated with the project objectives. The benefits of the approach are best articulated by Basili (1999, p. xiii):

> Writing goals allowed us to focus on the important issues. Defining questions allowed us to make the goals more specific and suggested the metrics that were relevant to the goals. The resulting GQM lattice allowed us to see the full relationship between goals and metrics, determine what goals and metrics were missing or inconsistent, and provide a context for interpreting the data after it was collected. It permitted us to maximise the set of goals for a particular data set and minimise the data required by recognizing where one metric could be substituted for another.

3 The Project Performance Framework

The project performance framework (PPDF) provides a formal approach to help practice in the elicitation and development of IS project objectives and measures that represent the values of the multiple project stakeholders. The PPDF involves an iterative series of steps during the life of the project (i.e. design to product delivery) to help identify objectives that the stakeholders want *of* and *from* the project, and measures that are directly associated with these objectives. The identification of stakeholders from design to delivery of the product is necessary to better account for the values essential to the perception of success (Fig. 2 in Chap. 5). The activities to elicit and structure represent the formative stages of designing the project criteria and measures. To accomplish its purpose, the PPDF relies on the principles of established methodologies. The VFT and GQM were chosen partly because of their high level of cohesiveness i.e. the VFT focuses on a structured approach to elicit objectives whereas the GQM begins in measure generation with the identification of objectives. Within our context, where one begins another ends i.e. in the elicitation of the objectives allows for the analysis to be performed for generation of measures. Additionally, the techniques support and facilitate stakeholders' participation which is an important ingredient of the PPDF. The development of the project's knowledge base or learning is also implicit in the structure as the development of competencies that are transferable to other projects can enhance the performance of the organizational projects (Fig. 2).

3.1 Identify Project Stakeholders

We propose a strategy for identifying stakeholders through the examination of the individuals or groups that are involved or may be impacted by the dimensions of the project performance (project process, PM process and product), and is influenced by the context of the specific project. The steps:

- Identify the individuals or groups that play an active role in the project. Resources for this step include: organization chart and roles and responsibility matrix can provide guidelines to this activity.
- Identify individuals or groups that are involved or are impacted by the **project process**. Within each the phase of project there may be a unique set of stakeholders that are relevant. As the project evolves the stakeholders may also grow.
- Identify individuals or groups that are involved or are impacted by the **project management process**. There are many persons that may be included in this net outside of the project manager and team
- Identify individuals or groups that are involved or are impacted by the **product**. For example the client or end-users are some immediate stakeholders, however depending on the actual product or outcome this may include other demographical groups.

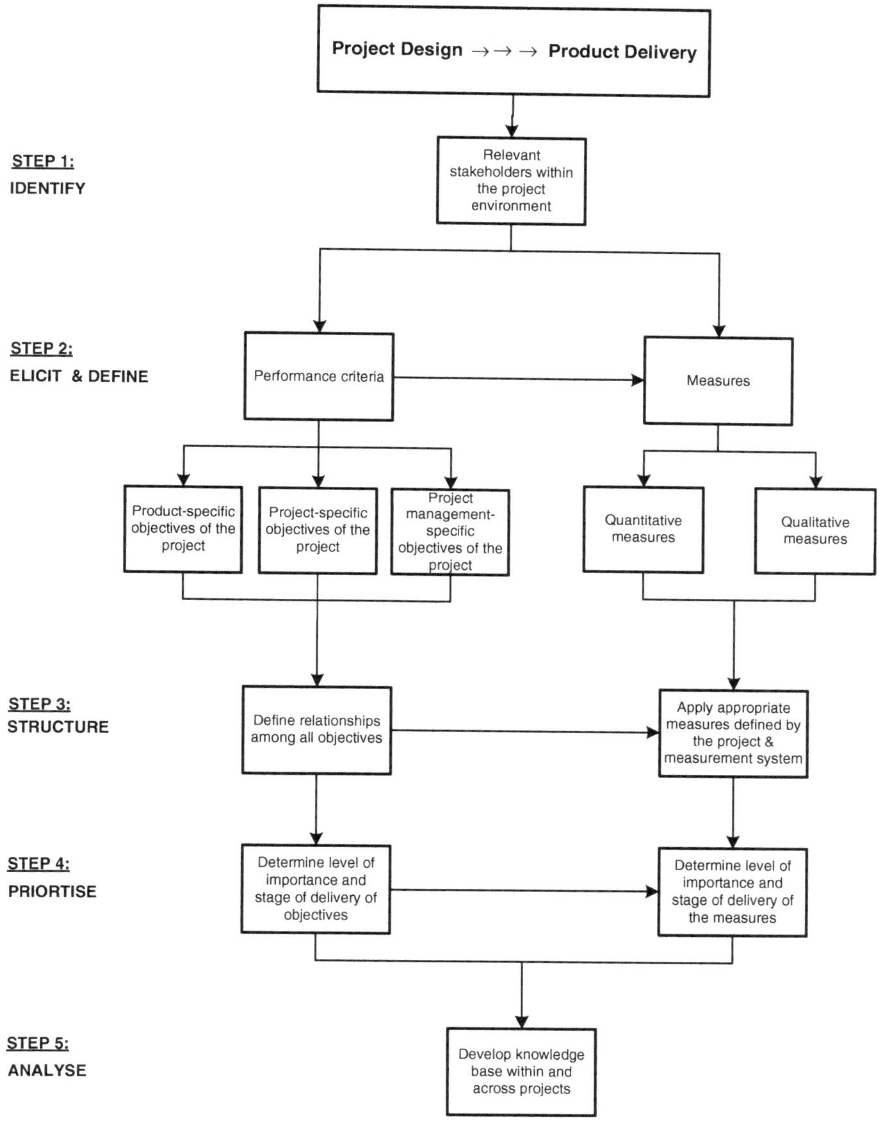

Fig. 2 PPDF framework

- Reduce the common set of stakeholders. It is expected that there may be some overlap so checking for and eliminating redundancies are necessary components of this step.
- Prioritise stakeholders based on stakeholder influence and impact in the project. It will be difficult to eliminate the subjective influence of this step; however, the key is to not diminish the views of the other stakeholders.

- Categorise stakeholders based on specific groupings, e.g. dimensions of project performance, project life cycle or influence.
- Review and refine stakeholders' list through-out the project.

3.2 Identify and Structure Project Objectives

This step adapts some of the steps of the VFT method:

1. Develop the list of objectives. This involves engaging the stakeholders and eliciting the set of objectives from each stakeholder based on the decision context or project purpose. These are placed into a common form which is primarily done to address different descriptions for similar set of ideas or objectives.
2. Structure objectives into *fundamental objectives* and *means objectives.*
3. Organise the objectives. The set of project objectives are further classified into the components of the dimensions of project performance i.e. PM, project and product objectives. This categorization further aids the decision-making process as there is improved understanding of the range of objectives and when they may be achieved combined with how the may influence each other.
4. Develop the stakeholders-objectives network. This extends the means-end objectives network by providing a visual connection between stakeholders and their fundamental objectives.
5. Review results. Review of the network in order to ensure completeness of the objectives and accuracy and completeness of the relationships. The main questions asked during this process were: did it make sense and did it accurately reflect the views of the stakeholders. Refinements are performed to address any missing links and consistencies among the project objectives.

3.3 Elicit and Define Project Measures

This step involves the following steps of the GQM method:

1. Formalise measurement goals.
2. Identify quantifiable questions.
3. Define the measures to be used.

3.4 Prioritization of Project Objectives

Several decision techniques are suitable in prioritizing the project objectives, some of which could be conflicting such as various techniques for ranking and scoring

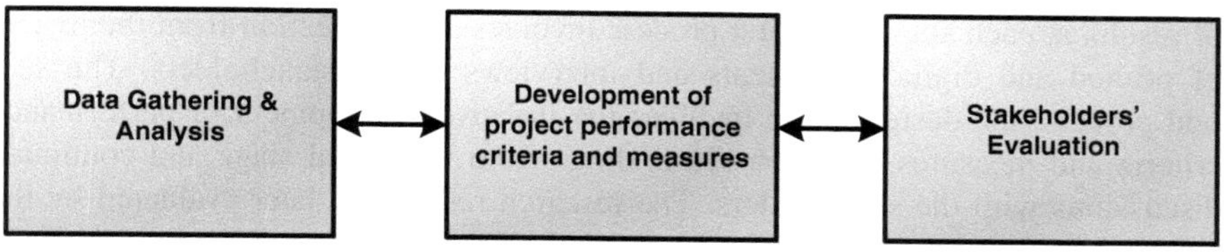

Fig. 3 Illustration and evaluation approach

[e.g. Simple direct ranking, Qualitative Discriminant Process (Bryson 1997)], and Multiple Criteria Decision Analysis (MDCA) techniques such as the AHP of (Saaty 1990) which is useful for resolving complex problems involving high stakes and diverse perceptions (Bhushan and Rai 2004). Alternately a Likert-like scale of 1–5 ranging from *low* to *high* priority, could also be used to further gauge the importance of the project objectives from the stakeholders. It should be noted that the PPDF provides flexibility in the choice of prioritization technique and does not advocate any one prioritization technique in this research.

4 Illustration and Evaluation of PPDF: Methodology

4.1 Overview on the Case Study Approach

The case study approach is recommended for the in-depth study of an artefact in the business environment (Hevner et al. 2004). We therefore used it to evaluate our artefact (i.e. the PPDF) in the context of a single case study. The case study approach embodies an account of past or current phenomenon drawn from multiple sources of evidence (Leonard-Barton 1990). It is further explained as *an empirical inquiry that investigates a contemporary phenomenon within its real-life context; when the boundaries between phenomenon and context are not clearly evident; and in which multiple sources of evidence are used* (Yin 1994, 2003). The approach allows researchers to better understand the nature of the organizational processes thereby providing deeper insights into the project contexts. For example, it provides us with the opportunity to understand how the PPDF fits into the real-life contexts (Darke et al. 1998) of the selected projects while providing us with important data for further refinements. Our use of the case study approach can also be considered to be a response to calls for more empirical research in project management (Winter et al. 2006a), with an emphasis on IS projects.

To facilitate the evaluation of the results from the data collected, the research strategy was appended to include three distinct stages of analysis: *data gathering and analysis*, *development* of the projects' performance criteria and measures, and the *evaluation* of these models by the stakeholders (Fig. 3).

The approach facilitated iterative interactions with the different stakeholders through data gathering, development of criteria and measures and evaluation of results at each stage. The first process involves data collection from the review of printed and digital documents and interviews of the stakeholders. The second process, the design phase involves the iterative development of performance criteria and measures based on the findings from the initial stage and continued discussions with the stakeholders. The research results are later evaluated by the stakeholders in terms of the usefulness of the model itself and the results of the application of the PPDF, i.e. derived objectives and measures.

4.2 Data Collection

The study is intended to accomplish several key objectives including to illustrate the steps of the PPDF and demonstrate its ability to provide aid to current IS project management processes and to gather stakeholders' perspectives on how the model has helped them in viewing and analyzing the project, and its performance. An initial questionnaire was distributed to the project owners to obtain background information on the organization and the project, including establishing the rationale for the project. Subsequently, face-to-face and telephone structured and unstructured interviews were used to gather perspectives. Three rounds of interviews supported by discussions with the stakeholders were performed. The first round was used to gather basic information about the company and the project and its objectives. The VFT and GQM techniques were used in subsequent rounds to elicit the facets of the projects that are important to the stakeholders and the appropriate measures and to refine the results. The duration of the interviews ranged from 40 to 60 min in the latter rounds, the initial set of interviews lasted between 60 and 90 min. As the development of the models of each project progressed, the project stakeholders were informed of the results, i.e. the objectives, measures and relationships, and interactive discussions were done to gather their opinions on its accuracy and completion and to revise the information where necessary. Meeting notes were taken during all the interviews and discussions. Additionally, records were reviewed at the beginning of the investigations and included project planning documents and company brochures and websites to provide supplemental background into the company and the projects under review. Table 2 shows the data collection summary of the resources used in the research. Several of the stakeholders were able to provide perspectives on different groups due to expertise and/or roles played in project. Based on factors such as individual's work experience, size of project team, individuals were able to "wear many hats" and provide additional insights into several perspectives.

4.3 Reliability and Validity of the Research

The principles of case study research were followed to enhance the rigor of our study. Yin (2003) proposed that the incorporation of some clearly outlined

Table 2 Data collection summary

Project	Interviews (# of participants)	Stakeholders perspectives considered
Process quality management development (PQMD)	• Director (2) • ISO Policy Consultant (1) • Staff (1)	• Executive management • Project sponsor[a] • Contractor • End-user

Notes/explanation:
[a] A director was the majority owner of the company and project sponsor for the project

principles will substantially increase the case study's quality. It was concluded that key principles to adhere to during the data collection effort include:

1. Multiple sources of evidence that converge on the same set of facts or findings. As evident in Table 2, multiple sources were used to collect information about the project. Several stakeholders were interviewed across the projects to understand their views on what is important in their project.
2. Maintenance of a case study database, distinct from the final case study report. Thus, notes from discussions and interviews are compiled and maintained separately from the final report to also provide a chain of evidence.
3. A chain of evidence. Similar to above, the results of investigations including companies' documents, notes are kept as evidence to support the research and its findings.

4.4 Illustration and Evaluation: Research Results

4.4.1 Preliminary Findings

Observations of the participants in their environment along with discussions revealed some findings pertinent to our research, including the current practice of identifying project objectives and evaluating the performance of their projects. There is a relative lack of attention or collaboration in identifying or developing project objectives. A project manager commented that "The respective team leaders are responsible for developing these documents and circulating to the rest of the team after completion…" It was noted that they relied on experience to identify the objectives which may cause inconsistency in the level of completeness depending on the team member involved. Across the organizations, there was a strong reliance on the traditional triple constraint methodology to assess performance by the practitioners, which conforms to others empirical studies (KPMG 2005; White and Fortune 2002). A participant for example, noted that it is difficult to "walk away from the [project management] bible". This implies in part that there is restrictive thinking in how the project is viewed and variables used to analyse performance. Notably, the organizations do not always develop documentation to support their projects. Several of the organizations did not create formal documents of their project goals, objectives or results of their progress. This can be explained by lack of commitment of negligence (Huber 1991) because all the case participants were experienced project professionals. One boasted that they "never create [formal] documents yet their projects were always on time". A director admitted this current limitation but did not provide a reason for the continued practice. While we are not suggesting that documentation of every aspect of the project is mandatory, we argue that continued practice can allow for missed learning opportunities.

Against this background, the PPDF intends to serve as a learning tool and to provide a formal approach to develop performance criteria and measures that are aligned to business and stakeholders needs. Thus, the case study serves the multi-purpose of illustrating the PPDF framework while repositioning the view of the projects by helping to better account for stakeholders' perspectives on the project objectives or criteria and measures. The study also indicates that the procedures of the PPDF helped in identifying additional criteria and measures that may not have been identified and tracked under different circumstances.

4.5 Project: Process Quality Management Development

4.5.1 Project Background

The Caribbean-based organization offers software and operational consulting services to businesses in various sectors such as mining, financial services and government. It was formed five (5) years ago and has a team of experienced full-time and contractual consultants and administrative staff managed by a team of directors. With a vision to expand its market base and create a lean and flexible organization it has embarked on a series of projects to achieve this mandate. A part of this mission is the achievement of ISO 9001 certification within the next 2–3 years. Thus, this exercise is seen as an important step in further legitimatizing the company and its services as one of the directors noted that they hope to become more competitive as a result of these new developments.

One of the projects involves the development of a quality management program. This project is initially slated to last 4–6 months with a major output being the organizational quality management methodologies, policies and procedures. The stakeholders of the project are categorised into project, project management and product stakeholders (c.f. project performance) and include the company's team (director and staff) and an external policy consultant (c.f. Table 3). The project stakeholders had consensus in how the project would be evaluated, i.e. based on the completion of policies and procedures and adherence to the scope of ISO framework. The project sponsor added that "adherence to time and budget will not determine the success of the project as we will have the set of documents from which we can work even if those targets are not met". As per ISO 9001 standards

Table 3 PQMD project summary

Project name	Process quality management development (PQMD) project		
Purpose	To set-up organizational policies and procedures to help standardise service delivery		
Start date, duration	March 2008, 4–6 months (1st phase). The remaining activities are projected to last next 2–3 years		
Key stakeholders	*Stakeholder Perspectives*	*Project and PM Processes*	*Product Process*
	Project sponsor	X	*x*
	Executive management		*x*
	Contractor	X	*x*
	End-user		*x*
Initial set of project objectives	• To complete policy & methodology documents • To minimise time to completion • To minimise project costs		
Initial project performance measurement	• Standard methodology: scope, time and cost • Completion of the development of organizational policy + methodology suite of documents		

(ISO/IEC 2000), some of the requirements include a set of procedures that cover all key processes in the business and thereby guide this project include: keeping adequate records; monitoring organizational processes to ensure they are effective; checking output for defects, with appropriate corrective action where necessary; regularly reviewing individual processes and the quality system itself for effectiveness; and facilitating continual improvement

4.5.2 Applying the PPDF

At the time of the research, the organization was at the initiation phase of the project and admitted that it did not perform any formal planning of the project and as such there were no formal documentation, such as a project charter available for review. Notwithstanding, the participants were able to articulate parts of their vision on what they expect from the project, however, the issue of budgetary limits and timeframe were nevertheless vague. This scenario presented the researchers with the opportunity to utilise the PPDF to help the organization in identifying and structuring their objectives and measures of the project which provided aid in determining the performance of their activities. The decision context or overall objective of the project is to improve the quality of its consulting service.

Identify Stakeholders. The individuals that play an active role in the project include the external consultant, sponsor and staff. In applying the additional steps, an analysis of those involved or impacted by the project, PM and product dimensions included executive management, current and potential clients and auditors. The stakeholders identified were relatively small which may be indicative of the size of the project.

Elicit and structure project objectives. The stakeholders were asked questions such as "what values were important in the project", "what was their wish list for the project" and "what factors would constitute a successful project" from their perspectives. The results of the interviews provided an initial set of objectives which were later compiled, reviewed and reduced into a common form. The results show a maximum of two rounds of refinement to the stakeholders-objectives-network. This strategy was applied to all the cases.

The process produced a final list of 18 project objectives for the PQMD project. The relationships among the objectives were developed, and the process of organizing the project objectives into the project performance dimensions was then performed by analyzing the dimensions in which they were applicable (c.f. Table 4 and Fig. 4). The results show a significant portion being in the dimension of the project's product (e.g. *maximise revenue, extend market reach*) and other remaining in project (e.g. *develop methodology, implement methodology)* and PM (e.g. *minimise project costs, minimise time to delivery of project outputs)* dimensions. The fundamental objectives identified included to *maximise efficiency, maximise staff competencies, maximise record-keeping and maximise client satisfaction.* The sponsor explained that to *maximise efficiency* was critical as they wish to become more competitive and run a more efficient business: "We wish to improve the way we do things… we would like to improve our core processes, and quality of service especially our service engagement [to clients]". It was also noted that to *maximise staff competencies* in the developed methodologies, policies and procedures was a necessary component to the success of their efforts: "…knowing our methodologies will [help to] reduce rework" and "we want to build awareness of the policies and procedures, [and methodologies]". To *maximise record keeping* is a core part of the ISO and they saw this as a means to improving maintenance of their project records: "We want to improve our documentation of our projects." To develop a reputation for quality was also found to be an important part of the project: "Build up a reputation of quality service and output by satisfying our customer"

Identify and Elicit project measures. In eliciting the measures, the purpose of the objective is first identified, followed by the its issue, object or aim of the objective, and viewpoint through which the objective is seen (Caldiera 1994). The stakeholders were also asked "what characteristics would indicate an achievement" of the individual objectives, and "how would they measure the project objectives" for each of the objectives identified. The set of questions are then developed from which the metrics of viewing the objectives derived and the process was repeated for all the objectives. For example, the objective of *maximise staff competencies* is analysed by identifying purpose of the objective (i.e. attain), the issue (i.e. maximization), the object of the objectives (i.e. staff competencies), and the viewpoints through which the objectives may be viewed (i.e. director, staff). Questions include (1) *What are the characteristics that indicate or reflect staff competencies?* and (2) *Are formal assessments of staff competencies performed?* Based on the analysis some of the measures include (1) the number of training and the results of training assessments, (2) quality of performance i.e. completeness or comprehensiveness of project deliverables, and (3)

Table 4 PQMD project objectives and measures

Performance dimensions	Project objectives	Measures
Project management	1. Minimise time to deliver project outputs	• Time of delivery against agreed standards
	2. Maximise training	• # of training hours per period
	3. Minimise project costs	• Conformance to budget • Results of earned value analysis (EVA)
Project and Product	4. Maximise staff competencies	• Reduction in # of rework • Comprehensiveness or completeness of deliverables (perception of quality)
	5. Maximise positive audit results (internal)	• Results of audits conducted • Reduction of non-compliance issues identified
Project	6. Develop organizational policies and procedures	• Completion of procedures/methodologies • Completion of policy document
	7. Implement QM methodologies	• Results of staff feedback • # of references to methodologies (i.e. use of policies and methodologies) • Results of internal audit
	8. Maximise record keeping	• # of records created (electronic and printed) • Incremental increase in records created
Product	9. Maximise clients' satisfaction	• Results of customer satisfaction index survey • # of repeat clients • Incremental increase in # of projects per customer
	10. Develop quality reputation	• # of repeat clients • Results of customer satisfaction index survey
	11. Extend market reach	• # of new customers • # of new customers in new markets • Penetration of customers in segments
	12. Increase customer base	• # of new customers • Incremental increase in # of projects per customer
	13. Maximise client retention	• # of contracts per client
	14. Maximise efficiency	• Reduction of non-compliant audits • # of new contracts • Reduction in time to complete clients' services • Reduction in costs per project
	15. Maximise revenue	• % age increase in revenue • Average revenue per customer
	16. Maximise management reviews	• # of meetings • # of corrective actions • Results of corrective actions • Expected implementation versus actuals
	17. Attain ISO certification status	• Results of ISO audit • Achievement of certification

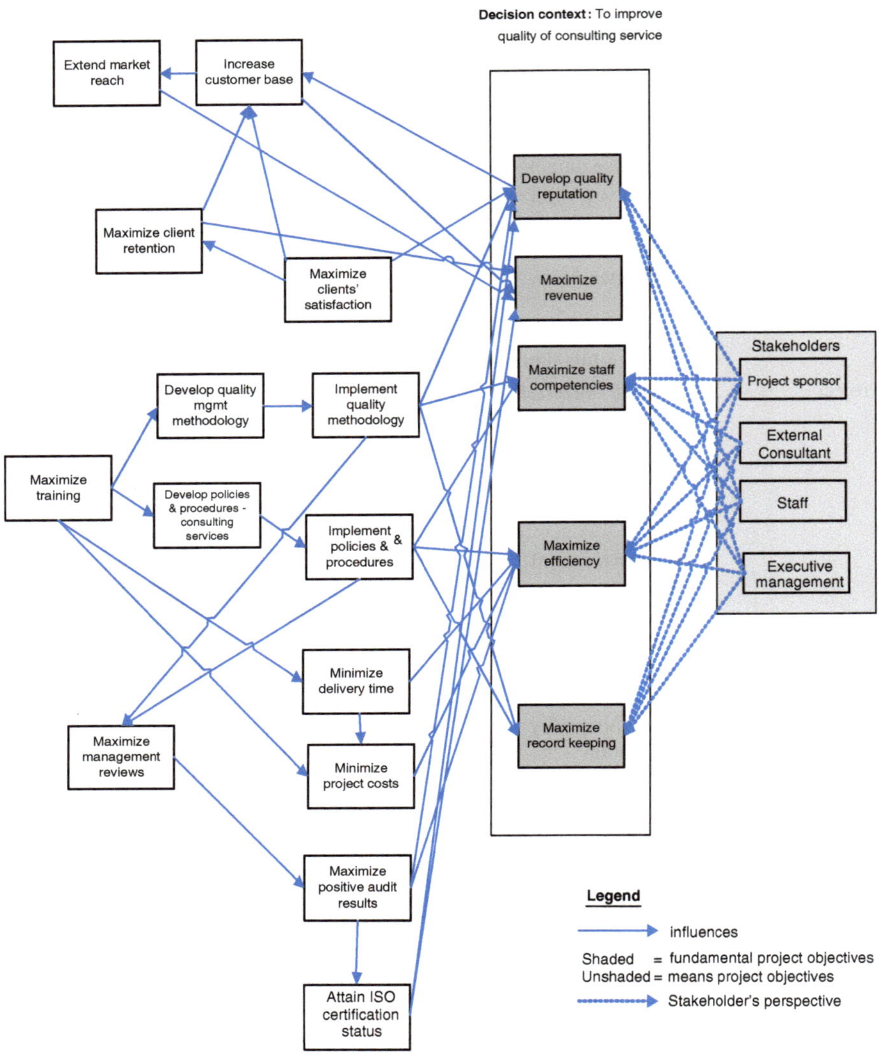

Fig. 4 PQMD project stakeholders-objectives network-2 rounds

the number of rework done for each deliverables. An analysis was done on the measures to ascertain the suitability of those identified to the stated objectives and to help determine if there were also any missing measures to enhance the comprehensiveness.

Prioritise objectives. The results of the prioritization revealed that they rated most of the objectives highly and the Average Priorities (AP). Objectives with the highest mean score of 4.5 included the development and implementation of organizational policies, increase revenue and development of a quality reputation. The objective to *extend market reach* received the lowest priority probably because this is seen as a longer term objective and is impacted by other exogenous influences.

4.5.3 Evaluation Perspectives of PPDF

The stakeholders, particularly, the sponsor indicated that the exercise and the results of the PPDF were helpful in understanding the purpose of the project. "Because we did not have a formal set of documents the charts [means-end network] helped me to think clearly about the project and what I wanted to achieve…" While the same can be said about tools such as project charter, the relationships among the objectives, the elicitation of measures associated with the objectives enhance the definition process and facilitate improved analysis for the remainder of the project. The ability to see how the objectives influenced each other was another major benefit they saw of the PPDF as under normal circumstances objectives were stated but the connections were not seen which sometimes caused disagreement among stakeholders, and impede the project process. The sponsor claimed that stakeholder interaction and buy-in was improved. "The process allows us to get buy-in from the stakeholders…they were able to see the goals of the project and contribute to the development of the project objectives." This can be attributed to frequent interaction among the stakeholders in eliciting important areas of the project and how they would characterise achievement of their objectives. This level of collaboration reinforces the importance of the project team's input, allows for early clarification of issues and the boundary or scope of the project. These events can lead to improved chance of meeting the stakeholders' expectations and the objectives of the project. These are some of the benefits underlined by the techniques and can become useful in other organizational projects. The issue of benefits is also important as the stakeholders did not fully consider the spectrum of measures available to assess their activities and often were considering only the most obvious objectives or the standard methodology in assessing the project as is evident from the initial project performance evaluation criteria (Table 4 in Chap. 5). For example, measures associated with learning in and from the project e.g. quality of service deliverables (time to completion, # of rework and customer satisfaction or perception to these deliverables) were not on the organization's radar initially. Therefore the model offers an additional analysis of their activities through the examination of the measures that are linked to each project objective. Concerns were initially expressed on how the management of the project scope could be handled once the inclusions of diverse perspectives and the "wish list" from the stakeholders were taken into consideration. However, based on our discussions and the result of the exercise it was reinforced that that the process will allow for issues to be brought to the fore and misunderstandings clarified and project scope clearly defined and understood at an earlier stage. Further, with the prioritization steps the stakeholders will be able to have an added point of clarification with the agreement on the project objectives that are deemed most important based on the context of individual projects.

5 Chapter Summary

The PPDF extends the application of several techniques, including the VFT and GQM to provide a guide in formally representing and accounting for the values of the project stakeholders. The research proposes that through the integration of the project process, project management and product objectives and formal methods to gather these, a more comprehensive set of guidelines can be developed to provide the view through which the performance of the project can be evaluated. The preliminary findings can provide additional benefits to project practitioners in the management of IS project processes and its stakeholders' expectations. The research further indicates that a strong agenda for success is laid when suitable performance criteria and measures are clear, unambiguous and representative of the stakeholders in the project. The PPDF is intended to engender tighter collaboration among stakeholders and better development of performance criteria and measures that can positively influence the measurement process.

One way to assure that clear and relevant objectives are gathered that are reflective of the individuals with a stake in the project is to have framework that uses robust technique(s) to accomplish this task. This is one of the critical goals of this paper. Our investigations found that the alignment of objectives and measures are important in having an efficient evaluation process. However, we also found that before this can occur, one also needs to ensure that the inputs into the process are accurate and reflective of the project and its participants. Therefore, the PPDF is established which attempts to address current gaps in literature.

A single study was used to illustrate and evaluate the utility of PPDF. The research underscored some of the challenges organisations have in identifying their project objectives and measures and documenting this project knowledge. The participants found that the method had utility for their project and environment, a key evaluation criterion for design science research (Hevner et al. 2004). Additionally the participants agreed that the application of PPDF helped in clarifying their project scope, manage stakeholders' expectations, and support stakeholders' involvement. These are important contributions of the research since multiple studies (Ewusi-Mensah 1997) have shown that these issues contribute to project failure.

Acknowledgments Material in this chapter previously appeared in: "PPDF: An Approach for Developing Performance Criteria and Measures for Information Systems (IS) Projects", *International Journal of Production Economics 124:1*, 272–292 (2010).

References

Agarwal N, Rathod U (2006) Defining 'success' for software projects: an exploratory revelation. Int J Proj Manage 24(4):358–370

Ala-Risku T, Kärkkäinen M (2006) Material delivery problems in construction projects: a possible solution. Int J Prod Econ 104(1):19–29

Alleman GB (2005) Agile project management for it projects. In: Y-HK, Carayannis EG, Anbari FT (eds) The story of managing projects: an interdisciplinary approach. Greenwood Press, New York, pp 373

APM (2006) The apm body of knowledge, 5th edn. The Association of Project Management
Arshinder, Kanda A, Deshmukh SG (2008) Supply chain coordination: Perspectives, empirical studies and research directions. Int J Prod Econ 115(2):316–335
Atkinson R (1999) Project management: Cost, time and quality, two best guesses and a phenomenon, its time to accept other success criteria. Int J Proj Manage 17(6):337–342
Atkinson R, Crawford L, Ward S (2006) Fundamental uncertainties in projects and the scope of project management. Int J Proj Manage 24:687–698
Azzone G, Masella C, Bertele U (1991) Design of performance measures for time-based companies. Int J Oper Prod Manage 11(3):77–85
Baccarini D (1999) The logical framework method for defining project success. Proj Manage J 30(4):25–32
Barclay C (2008) Towards an integrated measurement of is project performance: The project performance scorecard. Inf Syst Front 10:331–345
Barclay C, Osei-Bryson KM (2008) The project objectives measurement model (POMM): An alternative view to information systems project measurement. Electron J Inf Syst Eval 11(3):139–154
Basili VR (1999) Foreword. In: Solingen R, Berghout E (eds) The goal/question/metric method, a practical method for quality improvement of software development. McGraw Hill, London
Basili VR, Weiss DM (1984) A methodology for collecting valid software engineering data. IEEE Trans Softw Eng 10(6):728–738
Belassi W, Tukel OI (1996) A new framework for determining critical success/failure factors in projects. Int J Proj Manage 14(3):141–151
Bhushan N, Rai K (2004) Strategic decision making: applying the analytic hierarchy process. Springer, London
Bourne L, Walker DHT (2006) Using a visualising tool to study stakeholder influence—two australian examples. Proj Manage J 37(1):5–21
Bryde DJ, Robinson L (2005) Client versus contractor perspectives on project success criteria. Int J Proj Manage 23(8):622–629
Brynjolfsson E (1993) The productivity paradox of information technology. Commun ACM 36(12):67–77
Bryson N (1997) Supporting consensus formation in group support systems using the qualitative discriminant process. Interface between information systems and operations research. Ann Oper Res 71:75–91
Cannon-Bowers JA, Salas E, Converse S (1993) Shared mental models in expert team decision making. In: Castellan JNJ (ed) Individual and group decision making: current issues. Lawrence Erlbaum, Hillsdale, pp 221–246
Caldiera VRBG, Rombach HD (1994) The goal question metric approach. Encycl Softw Eng 2:528–532.
Chen SC, Dhillon GS (2003) Interpreting dimensions of consumer trust in e-commerce. J Inf Technol Manage 4(2–3):303–318
CMMI (2002) Capability maturity model® integration (cmmism), version 1.1, cmu/sei-2002-tr-011
Cohen DJ, Graham RJ (2001) The project manager's mba: How to translate project decisions into business success. Wiley, London
Collins A, Baccarini D (2004) Project success—a survey. J Constr Res 5(2):211–231
Cooke-Davies T (2002) The "real" success factors on projects. Int J Proj Manage 20(3):185–190
Darke P, Shanks G, Broadbent M (1998) Sucessfully completing case study research: Combining rigor, relevance and pragmatism. Inf Syst J 8:273–289
Dhillon G, Torkzadeh G (2006) Value-focused assessment of information system security in organizations. Inf Syst J 16:293–314
Dixon JR, Nanni AJ, Vollmann TE (1990) New performance challenge: Measuring operations for world-class competition (irwin/apics series in production management). McGraw-Hill Professional Publishing, Homewood
Drevin L, Kruger HA, Steyn T (2007) Value-focused assessment of ict security awareness in an academic environment. Comput & Secur 26(1):36–43

Engwall M (2003) No project is an island: Linking projects to history and context. Res Policy 32:789–808
Esteves JM, Pastor J, Casanovas J (2003) A goal/question/metric research proposal to monitor user involvement and participation in ERP implementation projects. Paper presented at the information resources management association conference (IRMA), Philadelphia, USA
Ewusi-Mensah K (1997) Critical issues in abandoned information systems development projects. Commun ACM 40(9):71–80
Fitzgerald G (1998) Evaluating information systems project: a multidimensional approach. J Inf Technol 13(1):15–27
Folan P, Browne J (2005) A review of performance measurement: towards performance management. Comput Ind 56:663–680
Fortune J, White D (2006) Framing of project critical success factors by a systems model. Int J Proj Manage 24(1):53–65
Freeman M, Beale P (1992) Measuring project success. Proj Manage J 23(1):8–17
Glass RL (2005) IT failure rates—70% or 10–15%?. Software IEEE 22(3):112–111
Globerson S (1985) Issues in developing a performance criteria system for an organisation. Int J Prod Res 23(4):639–646
Hartman F, Ashrafi RA (2002) Project management in the lnformation systems and lnformation technologies industries. Proj Manage J 33(3):5–15
Hevner AR, March ST, Park J, Ram S (2004) Design science in information systems research. MIS Q 28(1):75–105
Huber G (1991) Organizational learning: the contributing processes and the literatures. Organ Sci 2(1):88–115
ISO/IEC (2000) Quality management systems—requirements: international organization for standardization
Kaplan RS, Norton DP (1992) The balanced scorecard: Measures that drive performance. Harvard Bus Rev 70(1):71–79
Katzenbach JR, Smith DK (2003) The wisdom of teams: Creating the high-performance organization. Harper Collins, New York
Keeney RL (1992) Value-focused thinking: a path to creative decision making. Harvard University Press, Cambridge
Keeney RL (1996) Value-focused thinking: identifying decision opportunities and creating alternatives. Eur J Oper Res 92:537–549
KPMG (2005) Global it project management survey: KPMG Information Risk Management
Kumar RL (2003) Understanding the value of information technology enabled responsiveness. Electron J Inf Syst Eval, EJISE 1(1)
Leonard-Barton D (1990) A dual methodology for case studies: synergistic use of a longitudinal single site with replicated multiple sites. Organ Sci 1(3):248–266
Lim CS, Mohamed MZ (1999) Criteria of project success: an exploratory re-examination. Int J Proj Manage 17(4):243–248
Lyytinen K, Hirschheim R (1988) Information systems failures—a survey and classification of the empirical literature. In: Zorkoczy P (ed) Oxford surveys in information technology. Oxford University Press Inc, New York, pp 257–309
March ST, Smith GF (1995) Design and natural science research on information technology. Decis Support Syst 15:251–266
Martinsons M, Davison R, Tse D (1999) The balanced scorecard: a foundation for the strategic management of information systems. Decis Support Syst 25:71–88
Melville N, Kraemer K, Gurbaxani V (2004) Review: information technology and organisational performance: An integrative model of it business value. MIS Q 28(2):283–322
Milis K, Mercken R (2002) Success factors regarding the implementation of ict investment projects. Int J Prod Econ 80(1):105–117
Morris P (2004) Current trends in project and programme management. Association for Project Management (APM), High Wycombe
Nelson RR (2005) Project retrospectives: evaluating project success, failure and everything in between. MIS Q Executive 4(3):361–372

Niebecker K, Eager D, Kubitza K (2008) Improving cross-company project management performance with a collaborative project scorecard. Int J Manag Proj Bus 1(3):368–386

PMI (2004) A guide to project management body of knowledge, 3rd edn. Project Management Institute, Newtown Square

Pouloudi A, Whitley EA (1997) Stakeholder identification in inter-organizational systems: Gaining insights for drug use management systems. Eur J Inf Syst 6:1–14

Saaty TL (1990) How to make a decision: the analytic hierarchy process. Eur J Oper Res 48(1):9–26

Sharp H, Finkelstein A, Galal G (1999) Stakeholder identification in the requirements engineering process. Paper presented at the 10th international workshop on database & expert systems applications

Shenhar AJ, Dvir D, Levy O, Maltz AC (2001) Project success: a multi-dimensional strategic perspective. Long Range Plan 34:699–725

Shenhar AJ, Tisler A, Dvir D, Lipostevsky S, Lechler T (2002) Refining the search for project success factors: a multivariate typographical approach. R & D Manage 32(2):111–126

Siau K, Sheng H, Nah FF-H (2004) The value of mobile commerce to customers. Paper presented at the third annual workshop on HCI research in MIS, Washington, D.C.

Stewart RA (2008) A framework for the life cycle management of information technology projects: ProjectIT. Int J Proj Manage 26:203–212

Thamhain H (1994) Designing modern project management systems for a radically changing world. Proj Manage J 25(4):6

The Standish Group (2001) Extreme chaos

The Standish Group (2004) 2004 third quarter research report: The standish group international

Wateridge J (1998) How can is/it projects be measured for success? Int J Proj Manage 16(1):59–63

Wegelius-Lehtonen T (2001) Performance measurement in construction logistics. Int J Prod Econ 69(1):107–116

White D, Fortune J (2002) Current practice in project management—an empirical study. Int J Proj Manage 20:1–11

Winter M, Smith C, Cooke-Davies T, Cicmil S (2006a) The importance of 'process' in rethinking project management: the story of a uk government-funded research network. Int J Proj Manage 24:650–662

Winter M, Smith C, Morris P, Cicmil S (2006b) Directions for future research in project management: the main findings of a uk government-funded research network. Int J Proj Manage 24:638–649

Yin RK (1994) Case study research: design and methods, 2nd edn. Sage Publishing, Beverly Hills

Yin RK (2003) Case study research: designs and methods, 3rd edn. Sage Publishing, Beverly Hills

About the Authors

Kweku-Muata Osei-Bryson is Professor of information systems at Virginia Commonwealth University in Richmond, VA, where he also served as the coordinator of the IS PhD program during 2001–2003. Previously, he was Professor of information systems and decision sciences at Howard University in Washington, DC. He has also worked as an information systems practitioner in industry and government. He holds a PhD in applied mathematics (management science and information systems) from the University of Maryland at College Park.

His research areas include data mining, knowledge management, IS security, e-Commerce, IT for development, database management, decision support systems, IS outsourcing, multi-criteria decision making. He has published in various leading journals including *Information Systems Journal, Knowledge Management Research & Practice, Decision Support Systems, Information Sciences, European Journal of Information Systems, Expert Systems with Applications, Information Systems Frontiers, Information & Management, Journal of the Association for Information Systems, Journal of Information Technology for Development, Journal of Database Management, Expert Systems with Applications, Computers & Operations Research, Journal of the Operational Research Society, and the European Journal of Operational Research*. He serves as an Associate Editor of the INFORMS Journal on Computing, as a member of the Editorial Boards of the Computers & Operations Research journal and the *Journal of Information Technology for Development* and as a member of the International Advisory Board of the Journal of the Operational Research Society.

Ojelanki Ngwenyama is Professor and Director of the Institute for Innovation and Technology Management at the Ted Rogers School of Management, Ryerson University; Visiting Research Professor in the Faculty of Commerce, University of Cape Town; and Guest Professor in School of Information Science, Computer and Electrical Engineering at University of Halmstad, Sweden. Ojelanki holds a PhD in computer science from the Thomas J. Watson School of Engineering, State University of New York at Binghamton, and an MBA from Syracuse University. In 2009, he

K.-M. Osei-Bryson and O. Ngwenyama (eds.), *Advances in Research Methods for Information Systems Research*, Integrated Series in Information Systems 34,
DOI: 10.1007/978-1-4614-9463-8, © Springer Science+Business Media New York 2014

was awarded D.Phil (honoris causa) from the Faculty of Engineering, University of Pretoria, South Africa, for international contributions to informatics research. In 2012, he was VELUX Visiting Research Professor at the Copenhagen Business School, Denmark. And in 2011, he was an Andrew Mellon Foundation Mentorship Professor at University of Cape Town, South Africa.His papers have appeared in journals such as *MIS Quarterly*, *European Journal of Operational Research*, *European Journal of Information Systems, IEEE Transactions on Engineering Management,* and *Information Systems Journal*. He has served on the Editorial Boards of MISQ, JAIS, Information Systems Journal, Information Technology and People, the Scandinavian Journal of Information System, and The Journal of Information Technology For Development. He has been a member of IFIP WG 8.2 since 1987.

Kwasi Amoako-Gyampah is Professor, Department of Information Systems and Supply Chain Management, Bryan School of Business and Economics, at the University of North Carolina at Greensboro, USA. He obtained his PhD in operations management from the University of Cincinnati. His research interests are in managing technology and innovation, operations strategy, and supply chain management. His research has been published in Journal of Operations Management, European Journal of Operational Research, International Journal of Production Economics, International Journal of Production Research, International Journal of Operations & Production Management, Information & Management, the Database for Advances in Information Systems, Information Systems Frontiers, IEEE Security & Privacy, OMEGA, and others.

Francis Kofi Andoh-Baidoo is Assistant Professor, Department of Computer Information Systems and Quantitative Methods, University of Texas–Pan American. He obtained his PhD in information systems from Virginia Commonwealth University, Richmond. His research areas include data mining, knowledge management, IS security, e-commerce, IT for development, database management, decision support systems, and ICT use in developing nations. He has published in journals such as Expert Systems with Applications, the Database for Advances in Information Systems, Information Systems Frontiers, IEEE Security & Privacy, Journal of Information Technology Theory and Application, Information Resources Management, International Journal of Services and Standards, and International Journal of Electronic Finance. He serves on the editorial board of several journals including the Journal of Information Technology Theory and Application, International Journal of Electronic Finance, and Journal of Information Technology Education.

Arlene Bailey is a Lecturer in the Department of Sociology, Psychology, and Social Work of The University of the West Indies (UWI) at Mona. She holds PhD and MSc degrees in information systems and a BSc in computer science, all from the University of the West Indies. Her research areas include information and communication technologies (ICTs) for development, virtual communities, and communities of practice. Her work adopts a critical focus on the use of ICTs toward achieving social and economic development in communities. Her papers have appeared in several journals and conferences on ICT and development, including the *Electronic Journal of Information Systems in Developing Countries*, *Journal of Information Technology for Development*, and *Telematics and Informatics*.

Corlane Barclay is a certified project management professional and currently lectures at the School of Computing and Information Technology, University of Technology Jamaica. She designed and successfully implemented the School's 1st wholly owned masters in information systems management with five (5) specializations in 2011. She previously worked in various capacities in industry and government. She holds a PhD in information systems from the University of the West Indies, Mona Campus. She is currently engaged in several projects aimed at improving collaboration and development within the Caribbean. She is now actively also pursuing a profession in law upon receiving a law (LLB) degree in 2012.

Her research interests include project management, knowledge management, information security, law, data mining, decision support systems, and IT for development. She has published in the following journals: *Project Management Journal, International Journal of Production Economics, and Information Systems Frontiers*. She serves as reviewer to *Information Technology for Development Journal, Information Systems Management, and Journal of Evaluation and Program Planning*.

Felix Bollou is a faculty member in the School of Information Technology and Computing, American University of Nigeria; Research Scientist at the Institute for Innovation and Technology Management of the Ted Rogers School of Management, Ryerson University, Canada; and member of the SAP University Alliance User group. Felix holds a PhD in informations system from the School of Commerce, University of Cape Town, South Africa. He holds an engineering degree in computer science and a DESS (masters) from the Institute of Business Management, University of Paris 1, Pantheon-Sorbonne, France. His research is in the areas of information systems and applied economics. His current research focuses on the impact of information and communication technology (ICT) on development in West African countries, service-oriented architecture, and business intelligence. His teaching interests are in system dynamics, business process design, system analysis and design, enterprise integration, and database management. Felix has over 9 years of experience in academia and over 15 years of experience in IT consultancy, project management, and application development in various industries in Africa, Europe, and North America.

Myung Ko is an Associate Professor in the Department of Information Systems and Cyber Security at the University of Texas at San Antonio (UTSA). She received her PhD from Virginia Commonwealth University. Her research interests include impact of IT on organization, data mining, and economics of security breach. Her work has been published in refereed journals such as Information & Management, Information Systems Journal, Decision Support Systems,

Communications of the Association for Information Systems, Information and Software Technology, Information Resources Management Journal, Journal of Information Technology Management, Journal of Information Technology Theory and Application (JITTA), and Information Technology & Management.

Sergey Samoilenko is an Associate Professor and the Chair of the Department of Computer Science of Averett University, Danville, Virginia. Sergey's current research interests include IT and productivity, data mining, and IT for development. He holds PhD and MS degrees in information systems from Virginia Commonwealth University and BS in industrial engineering from the Institute of Soviet Trade Technology.He has published in the European Journal of Operational Research, Expert Systems with Applications, Information Systems Frontiers, Journal of Global Information Technology Management, International Journal of Production Economics, among other journals, as well as in numerous conference proceedings.

Manoj Thomas is an Assistant Professor in the Department of Information Systems at Virginia Commonwealth University. He holds a PhD in information systems, MS in information systems, MBA, and bachelors in engineering. His primary research interests are knowledge engineering, emerging technologies, and ICT for developing countries. He rides and works on motorcycles when he wants to get away from the digital realm.

Yan Li is working toward a PhD in Information Systems at Virginia Commonwealth University. Her research interests include knowledge management and data mining (KDDM), text mining, decision analytics, decision support systems, multiple criteria decision analysis, semantic technology and ontologies, and data and information quality management. Her current research in progress focuses on exploring the synergies between information systems and decision analytics, as well as applying KDDM methods and techniques in cross-disciplinary areas. She has presented at international, national, and regional information system conferences on topics such as strategic information quality management, decision support systems, geographic information systems, data mining, information system development, and information technology for developing economies.

Index

K.-M. Osei-Bryson and O. Ngwenyama (eds.), *Advances in Research Methods for Information Systems Research*, Integrated Series in Information Systems 34,
DOI: 10.1007/978-1-4614-9463-8,

MIX
Papier aus verantwortungsvollen Quellen
Paper from responsible sources
FSC® C105338

If you have any concerns about our products,
you can contact us on
ProductSafety@springernature.com

In case Publisher is established outside the EU,
the EU authorized representative is:
Springer Nature Customer Service Center GmbH
Europaplatz 3, 69115 Heidelberg, Germany

Printed by Libri Plureos GmbH
in Hamburg, Germany